Electronics for Service Engineers

Electronics for Service Engineers

Joe Cieszynski
and
David Fox

Newnes
OXFORD AUCKLAND BOSTON JOHANNESBURG MELBOURNE NEW DELHI

Newnes
An imprint of Butterworth-Heinemann
Linacre House, Jordan Hill, Oxford OX2 8DP
225 Wildwood Avenue, Woburn, MA 01801-2041
A division of Reed Educational and Professional Publishing Ltd

 A Member of the Reed Elsevier plc Group

First published 1999
Transferred to digital printing 2004

British Library Cataloguing in Publication Data
A catalogue record for this book is available from the British Library

ISBN 0 7506 3476 6

Typeset by Jayvee, Trivandrum, India

Contents

Introduction

The City & Guilds 2240 Electronics Servicing Course has been a well-established route for those wishing to gain certification and knowledge for a career in the brown goods servicing sector for many years. Like most vocational qualifications it has now been superseded by a new National Vocational Qualification (NVQ) and for those in Scotland a Scottish Vocational Qualification (SVQ). An NVQ or SVQ will be awarded to those people who can demonstrate that they can competently carry out a range of servicing tasks. This is quite unlike the previous qualification which was awarded when successfully completing a course of study. There is no compulsion to register on a course and attend at a centre for examinations and practical tests since there are none. It is the standards laid down by the representatives from the industry that have to be matched by a candidate. In theory this should ensure that the standards are relevant and current to a particular sector. Assessment of competence to gain certification can be carried out by a number of methods and could be in a customer's home, on company premises or in a college with the only stipulation being that those carrying out assessment must be qualified assessors with relevant experience. A qualified assessor is a person who has gained a D32 and D33 award from the appropriate awarding body. For those with many years' experience (and a great deal of knowledge) demonstration of competence can be supported through a route which accredits prior learning (APL). Therefore, there are many ways by which an individual can gain an award. Currently, the range of products covered by this award includes the following:

- Home laundry (washing machines, tumble dryers)
- Refrigeration
- Cooking equipment (gas/electric/microwave)

- Small appliances (vacuums/hair dryers/heaters)
- Audio (music centres/tape decks/CD players)
- Television/video
- Multimedia equipment
- Commercial products

The award covers all areas of a service engineer's job and incorporates the technical, customer care and health and safety aspects. NVQs at both levels 2 and 3 are available and in each level there are a number of mandatory units and optional units to be undertaken. A candidate can therefore pick those that apply to their circumstances. Although the demonstration of competence is the method used for assessment, candidates must also be able to show that they possess not only the practical ability but the knowledge and understanding required in each element since it is sometimes impossible to separate them. Even though performance evidence may be obtained through the outcomes of work activity and how a particular work activity was carried out, direct observation, work sheets or witness testimony, there must also be evidence that a candidate has the knowledge and understanding as specified in each element. Evidence of knowledge and understanding can be proved by the use of oral questioning or written questions and answers. The aim of this book is to provide the basic knowledge of electrical and electronic topics that is common to the widest possible product group range identified and is not aimed at any specific product group. It should therefore be as suitable for the small appliance, the audio or the commercial product service engineer.

1

Safe practice in the workplace

Every year numerous accidents, both minor and major, occur in the workplace. Yet in many instances these accidents could have been avoided if individuals had followed basic health and safety procedures, which quite often are nothing more than plain common sense. All too often people are injured because they, or someone around them, have either become complacent, have made an error of judgement, have been reckless, or have cut corners.

In many countries employers and trade unions have laboured to improve working conditions in order that industrial injuries are cut to a minimum. For many years in the UK the various Factory Acts, enforced by factory inspectors, have served to provide codes of practice that are designed to protect not only the employers and employees, but also the general public.

The Health and Safety at Work Act 1974 (HASAWA) expands on the provisions of the Factory Acts, and provides a comprehensive legal framework designed to promote high standards of health and safety in the workplace.

Of course, an act is no use unless it is enforceable. The HASAWA is enforced by a nationwide team of inspectors who are appointed by the Health and Safety Executive. These inspectors are entitled to enter a premises at any reasonable time, collect any evidence they believe relevant in cases where a breach of the Act is suspected, and inspect any relevant documentation. Where an accident has occurred, the inspectors will investigate the conditions surrounding the incident, and the executive may take action and prosecute anyone whom they consider may have caused the accident through a breach of the Act.

In 1980 the Notification of Accidents and Dangerous Occurrences Regulations (NADOR) was introduced. These regulations require employers and the self-employed to notify the local Health and Safety Executive office (usually within the local environmental health department) of any serious accidents, injuries, or cases of disease. In 1986 these regulations were superseded by the Reporting of Injuries, Diseases and Dangerous Occurrences Regulations (RIDDOR).

The HASAWA aims to involve everyone in matters concerning health and safety, whether they be employers, employees, self-employed, site/building managers, equipment and material manufacturers, or the general public. The scope of the Act is very wide as it encompasses all working conditions and environments, i.e. office, manufacturing, engineering, construction.

The Act states clearly the responsibilities of both the employer and the employee. These are summarised below; however, this chapter will deal mainly with those areas that appertain to electronic engineering.

Employer responsibilities

Figure 1.1 summarises the responsibilities of the employer. Failure to provide any of these could lead to prosecution, especially if it were proven that an accident had occurred because one or more of these provisions had not been made.

Figure 1.1

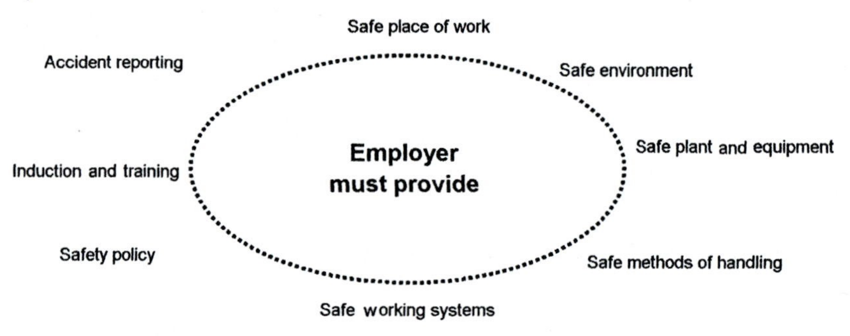

For the electronics servicing industry, providing a *safe place of work* generally means providing a building that is suitably maintained and is free from hazards.

A *safe environment* means having a temperature that is comfortable to work in, and an atmosphere that is clean not only from dust but also fumes, odours, fungi, etc. It also includes provision of washing, sanitation and first aid facilities. The reference to toxic fumes has become more applicable to electronic servicing workshops with the introduction in 1994 of the Control of Substances Hazardous to Health regulations (COSHH). Although these regulations cover a wide range of substances, in electronics it has altered the way large-scale soldering/de-soldering is performed. For many years it was acceptable to solder with a hand-held iron and simply duck out of the way of the high lead content smoke. Under COSHH the soldering equipment must have an extraction facility to prevent the atmosphere in the room from becoming toxic. In practice many inspectors look at this realistically, and do not insist that a small workshop with occasional soldering taking place should have extraction equipment. However, any premise where constant soldering takes place must be suitably equipped.

The COSHH regulations also affect the disposal of batteries. Throwing batteries into a waste bin means that they will more than likely end up on an open cast refuse tip where the toxic chemicals will eventually seep into the earth. All employers must make provision for correct disposal methods, which generally involve a plastic container specifically designated for battery disposal. This container can be emptied separately by the local authority cleansing department (for a fee), or in the case of lead/acid batteries, arrangements can be made with a recycling company for them to be collected.

Although engineers are usually expected to provide their own tools, the employer must ensure that all *plant and equipment* used at the place of work are safe. For a workshop, plant could be interpreted as benches, storage racking, trolleys, etc. Equipment would be test equipment, power tools, etc. Benches must be well maintained and at the correct height, whilst test equipment is maintained and PAT tested annually. All power tools must be safe, and have the correct safety guards fitted and working.

Employees must not be expected to move heavy items unaided. The employer must provide appropriate lifting and transportation equipment. In some cases it may be necessary for two people to be made available to move or transport an item.

Companies must devise *safe working systems,* which are procedures designed to minimise the risk of accidents. Once devised, these systems can be written up as a company *safety policy.* Policies such as this are not intended to be static, but dynamic. For example, when accidents occur, they will be investigated and the procedures modified if necessary. The employer must ensure that all employees are made aware of the safety policy, and any subsequent modifications to it. This should form part of an employee's initial *induction* and ongoing *training*. Examples of safe working systems are provision of protective clothing for certain tasks, provision of appropriate staffing levels for specific jobs, suitable warning notices of hazards or hazardous areas, and provision of clearly marked first aid facilities.

Under RIDDOR, employers' *accident reporting* involves informing their local health and safety inspectorate immediately following an accident resulting in death or serious injury. Incidents of a lesser nature may be reported within a certain length of time. The employer must also provide a reporting procedure for employees.

Employee responsibilities

These are summarised in Figure 1.2.

Figure 1.2

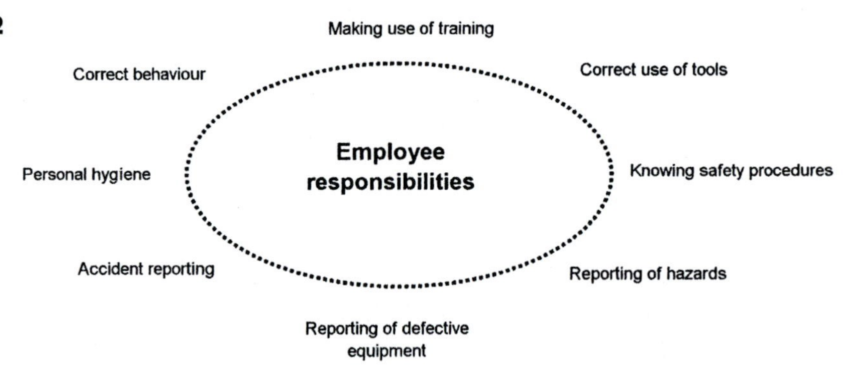

Making use of training. Safety policies can only be effective when each individual takes responsibility for their own actions, and ensure that they are observing safe working practice. All engineers are taught safe practice, and they must be aware that they may be held responsible for any accidents that occur should they fail to adhere to these practices. Such practices include *correct use of tools* and equipment, isolation of equipment from the mains voltage supply when carrying out certain jobs, not working whilst under the influence of alcohol or drugs (including certain prescribed drugs), wearing appropriate clothing, and wearing safety clothing where necessary.

Each employee must make it their business to familiarise themselves with the company safety policy, and to *know the safety procedures*.

Hazards and hazardous situations can arise at any time, and employees must make it their business to *report any hazards* to the appropriate person immediately. A hazard can be anything that could cause an accident, e.g. a loose carpet, defective storage racking, a blocked fire exit. Such things are so commonplace they can become a fact of life and hence ignored, but they are all accidents waiting to happen and should be reported so that they can be rectified.

Equipment and tools can be in use all day and every day, and this constant wear and tear often results in faults and defects that present potential hazards. Frayed mains leads, loose mains plug connections, broken casing on power tools; all of these *defects must be reported.* In a larger company there should be a reporting procedure where the fault would be reported in writing, and the employee finding the defect would attach an official warning label to the tool or equipment. However, even within a smaller company a defective item should be labelled to warn other employees of the defect.

All *accidents must be reported* in writing. Where a minor injury has occurred, the employee concerned must fill in details of the injury, and how it occurred, in the accident reporting book provided by the company. Where the injury is of a more serious nature and the employee is perhaps unable to give details, statements must be taken from any witnesses, and the employer must report the incident to the Health and Safety Executive office.

The employer is required to provide sanitation and cleaning facilities, but it is the employees' responsibility to make use of them. *Personal hygiene* in the workplace is very important if one is to avoid illness, skin irritations, etc.

Practical joking, horseplay, or an over-casual attitude has no place in a workshop or any other place of work. All employees are required to *behave appropriately,* and may be held criminally responsible for any accidents resulting from incorrect behaviour.

On-site servicing

It is relatively easy to create a safe environment within a designated workshop, but electronics servicing engineers are frequently required to service equipment on site, thus introducing all kinds of safety issues. For example, live circuits may need to be exposed in the vicinity of the general public, and in a domestic situation this may well include children. In such circumstances, how can an engineer be certain that he can control the environment? What might a child get up to whilst the engineer is at his vehicle looking for a part?

Other hazards are faulty electrical wiring within the building that could render the equipment unsafe to work on, lack of mains isolation transformers (which will be discussed later in the chapter), confined working space, and insufficient light. All of these must be taken into consideration by the engineer because ultimately *the engineer is the one who is responsible for safety* during the servicing operation.

If there are any doubts about safety whilst carrying out a task on site, the engineer should cease working and make alternative arrangements such as have the item of equipment brought into the workshop, return to the workshop for a complete working panel or mechanism, or, if these options are not viable, make arrangements with the customer to service the equipment when a safe environment can be created.

Electrical safety

Before looking at means of prevention of electric shock it is necessary to examine the effects of electricity on the human body, and the causes of electric shock.

All muscles are controlled by the brain via very small electrical impulses. When an external electrical current of a value greater than that generated by the brain passes through the body, it effectively takes over control of the muscles. There are three major consequences of this.

First, a current passing through the heart (which is its self a muscle) may cause it to cease its pumping action, the commands from the brain being overwhelmed. The person thus suffers from a cardiac arrest.

The second consequence may be that a person becomes 'stuck' to the live conductor. This occurs when a person touches a live conductor with their hand, and the polarity is such that the signal to the muscles in the hand is effectively saying 'contract'. The hand now grabs the conductor and the person is unable to let go, even though they may try to do so; the small 'let go' command from the brain being totally swamped. Left in contact with the conductor, heat is generated and the person may receive severe burns, especially if the voltage is sufficiently high (>50 V).

The third consequence may be that, on touching the live conductor, the polarity of the current commands the muscles in the arm to reflex. The person appears to throw themselves from the conductor, although they have no control over this action. A danger with this is that there are no guarantees where the person will land, and serious physical injury may result.

In both the second and third cases, cardiac arrest could occur if the current takes a path through the heart, and the value of current is high enough.

Remember, as a general rule: **Current Kills; Voltage Burns.**

A current of less than half of one amp (electrical current will be discussed in detail in Chapter 5) is enough to kill a person, and many safety cut-outs such as those purchased in high street stores for use with domestic gardening tools, etc. are designed to trip at 30 mA (0.03 of one amp).

The greatest electrical shock hazard for electronics service engineers is when working on live equipment that has a high voltage dc source, especially if the current path is across the chest. High voltage dc supplies are found in equipment such as colour TV receivers, microwave ovens, and large audio power amplifiers.

It will be seen later that electrical current always travels from a power source, around a circuit, and back to the power source once again. It therefore follows that if a service engineer can take steps to prevent himself from making a circuit, he will not suffer electrocution. There are a number of ways that this can be achieved.

First of all there is the single handed approach to servicing which is illustrated in Figure 1.3.

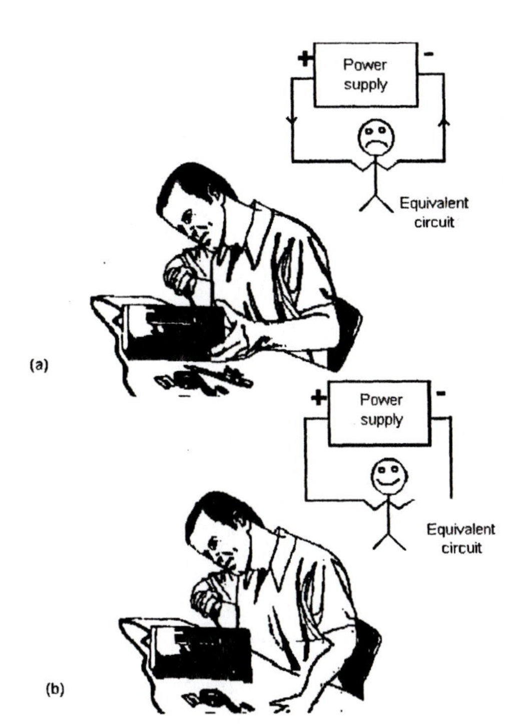

Figure 1.3

(a) Current flows from the positive supply (right hand) through the body and back to the negative supply (left hand).
(b) There is no circuit so current does not flow, provided that the bench top is an insulator

In Figure 1.3(a) the engineer has his left hand on the negative metal case (or chassis) of the equipment on which he is working. If the right hand were to slip and touch a high voltage contact, a current path would exist between his hands, taking a path straight across his chest and therefore the heart. It has already been shown that this is a very dangerous situation.

Figure 1.3(b) shows the engineer working with one hand away from the casing (it is on the bench; however, it could be placed on the plastic or wooden cabinet of the equipment). In this case there is no circuit and hence there can be no current flow.

Unfortunately safe working practice is not as simple as just working on live equipment with one hand. A serious shock hazard exists from the presence of the electrical earth path illustrated in Figure 1.4. All mains electrical supplies have an earth terminal which is connected to the neutral pole at the power station. This is included for safety purposes which shall be examined later. The normal current path is from the live pole (often called the *phase*; *P*), through the appliance, and back to the power source via the neutral conductor. The earth conductor does not normally carry any current.

Figure 1.4
Live, neutral and earth configuration in a domestic 230 V mains supply

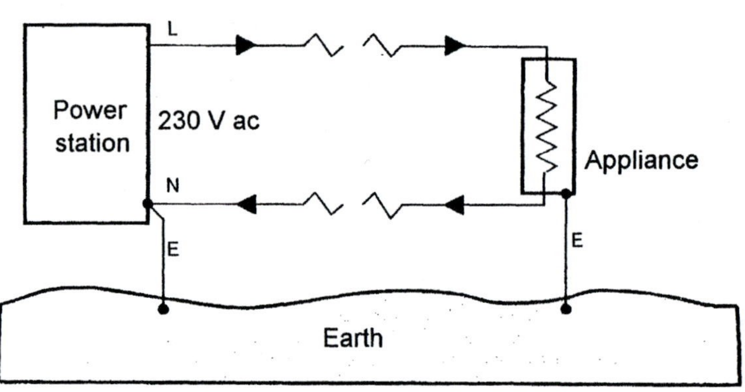

Now consider the situation shown in Figure 1.5. The engineer has come into contact with the live conductor. Because he is standing on the earth a current path now exists from the live, through his body, and back to the power source neutral via the earth; literally the ground on which he is standing.

The engineer does not have to come in direct contact with the live conductor. A similar shock hazard exists when working on certain types of equipment, notably colour televisions, where the chassis is at a half mains potential.

For the situation shown in Figure 1.5, the engineer could be protected by placing a rubber mat on the ground where he is working. Rubber, being an electrical insulator, would break the circuit and prevent an electric shock. However, a mat on its own is not satisfactory because the engineer might come into contact with the earth via another route. For example, by touching a radiator or other metal pipe, touching a solid masonry or stone structure, or perhaps through the earthed leads of the bench test equipment or soldering iron.

Figure 1.5
An electrical circuit exists between live and neutral via the earth path

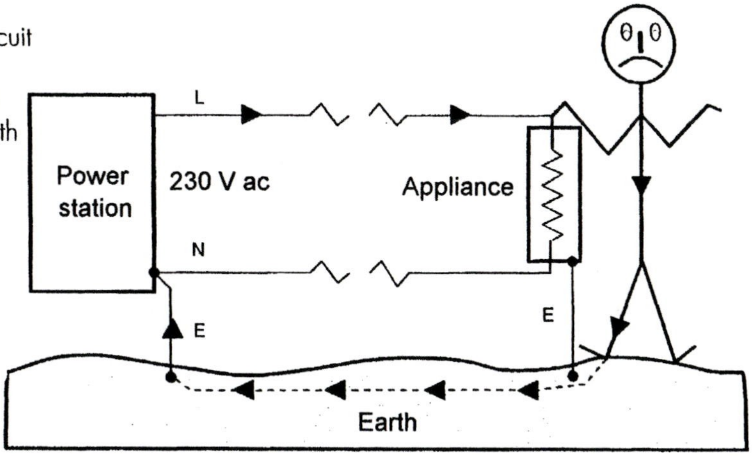

The most effective means of overcoming the type of shock hazard illustrated in Figure 1.5 is to use a *mains isolation transformer*. The principle is shown in Figure 1.6.

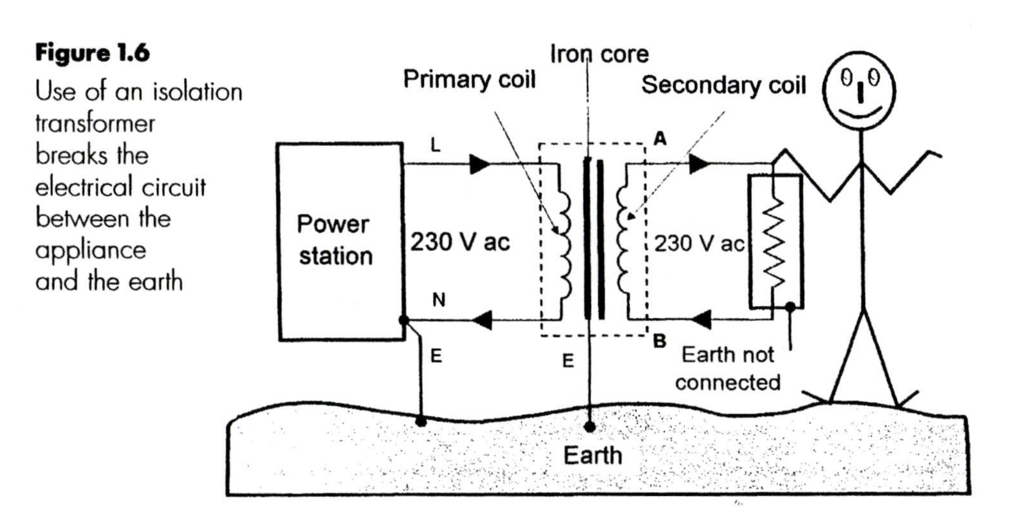

Figure 1.6
Use of an isolation transformer breaks the electrical circuit between the appliance and the earth

The transformer is used to break the earth path between the mains power source and the appliance. The operation of transformers will be looked at in Chapter 9; however, for now it can be considered to be a device that couples the 230 V input from the live and neutral to the appliance at points A and B using a magnetic field. There is no current path from the power station to the appliance.

Current from the 230 V mains supply flows from the live, through the transformer primary coil, and back through the neutral. This current sets up a magnetic field that induces a 230 V supply in the secondary coil. Current leaving the secondary coil at point A will be trying to return to the secondary coil point B, not the neutral. Should the engineer come into contact with the supply at point A, there will in theory be no current flow through his body because there is no circuit path back to point B. In practice a small current may flow due to leakage in the transformer; however, this *should* not be large enough to pose any serious threat.

It is a regulation that any mains powered equipment must be connected via a mains isolation transformer before it is opened up for service. This even applies when working away from the workshop, and portable isolation transformers are available for this purpose although they are not popular with engineers owing to their considerable weight. However, should any accident occur as a direct result of working without an isolation transformer, the health and safety inspectors may prosecute the person or persons responsible (if they are still alive!).

A problem frequently encountered with bench-based servicing is where the equipment has no connector attached to the mains flex because of the way it was disconnected from the supply when it was brought in for service. A simple device used to overcome this problem is the mains connector block, sometimes referred to as a 'safeblock'. This allows the live, neutral, and earth connections to be made using sprung terminals, the 230 V supply only being connected when the top cover is closed.

It is common practice to put a connector block on the output of the mains isolation transformer for the workbench. Although this means that the mains plug must be removed from every item of equipment that comes onto the bench, this definite action eliminates the chance of an engineer accidentally connecting the equipment to the non-isolated mains supply.

Electrical protection devices

The most common device for protecting electrical equipment is the fuse. This is a piece of wire with such properties that it heats up and melts if a current greater than a predetermined amount flows through it. When it melts, the electrical circuit is broken and current flow stops. Fuse ratings are very important and when replacements are fitted it is essential that the correct type and rating (in amps) is used, otherwise the fuse may not melt in the event of a further fault in the equipment, resulting in a fire or electrical shock.

There are many types and construction of fuse, each having its specific applications. Two common cartridge types are illustrated in Figure 1.7(a) and 1.7(b). Fuses will be discussed in more detail in Chapter 11.

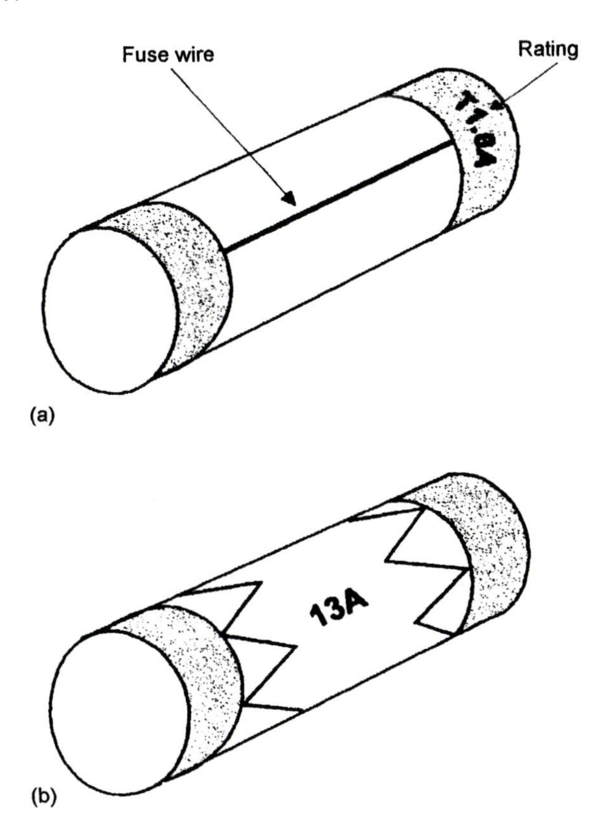

Figure 1.7

Common fuse types:
(a) Clear glass fuse. Usually 20 mm or 33 mm long.
(b) 25 mm 'plug top' fuse

It was stated previously that the earth connection (illustrated in Figure 1.4) is included as a safety feature. Figure 1.8 shows how the earth circuit path performs this safety function when a fuse is included.

Under normal conditions the current flow in the kettle circuit is from live to neutral. There is no current through the earth.

Now consider the fault condition shown in Figure 1.8(b) where, due to a fault, an electrical connection has occurred between the element and the metal container. If, as shown in this illustration, there is no earth connection, the kettle would still function correctly because there is still a circuit between live and neutral. However, should someone touch the container, a current path would exist from live, through the fault connection to the container, through the person, and back to the supply source.

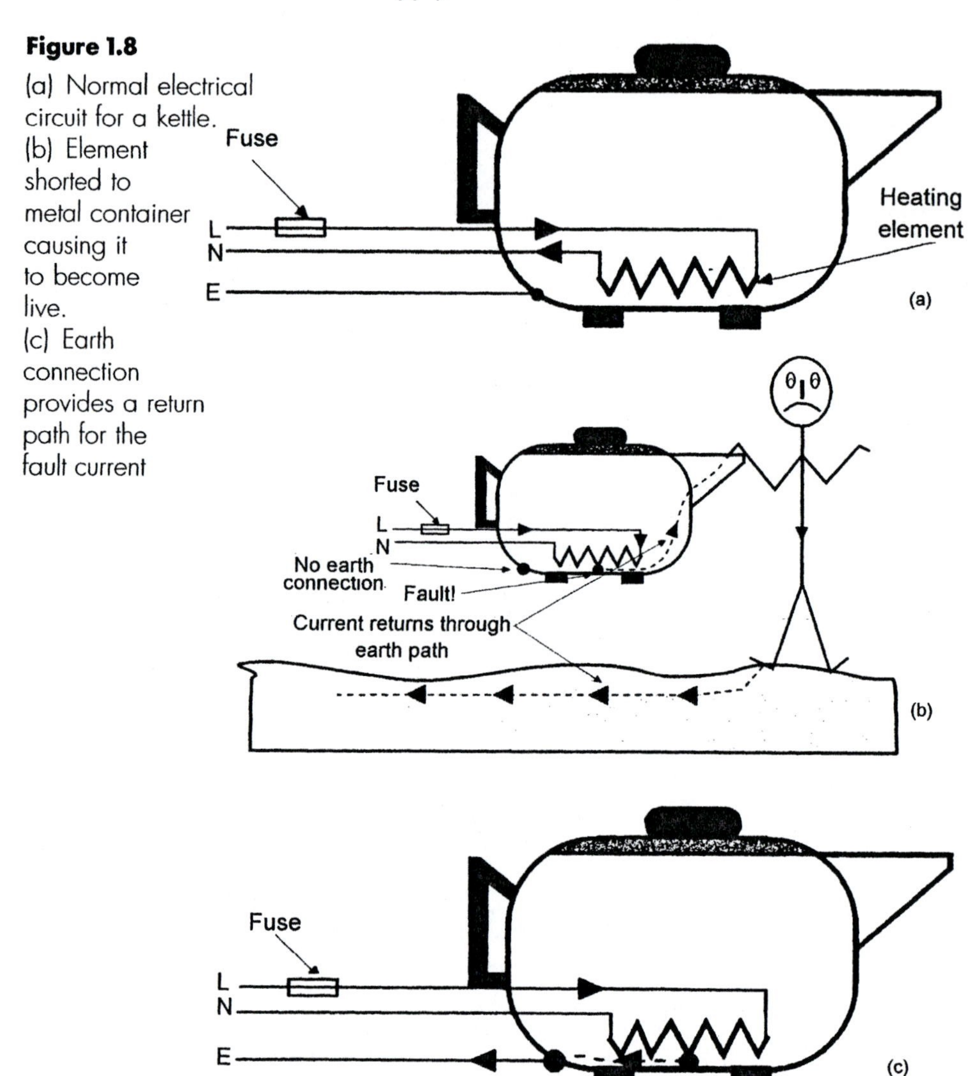

Figure 1.8
(a) Normal electrical circuit for a kettle.
(b) Element shorted to metal container causing it to become live.
(c) Earth connection provides a return path for the fault current

The inclusion of the earth connection in Figure 1.8(c) would mean that as soon as a fault occurs, a current will flow from the live to the earth and, if the fuse rating is correct, the excess current should cause the fuse wire to melt.

Recent years have seen the inclusion of earth current trip devices in the mains electrical supply. These are fast acting cut-outs that activate (usually) when a current of 30 mA (0.03 A) or more passes through the earth conductor for a period of no more than 40 msec (0.04 of one second). The principle of operation is simple. Taking the situation shown in Figure 1.8(a), all of the current flowing through the live conductor passes back out through the neutral conductor. However, in Figure 1.8(b) the amount of current passing through the neutral is less than that entering through the live because some current is returning through the earth (via the person).

An earth current trip device monitors the current in the live and neutral conductors. As soon as an imbalance of 30 mA occurs the device trips, breaking both the live and neutral connections to the equipment.

Earth current trips are usually called an *Earth Leakage Current Breaker (ELCB)* or a *Residual Current Device (RCD)*.

All workshops where electronic equipment connected to the mains voltage supply will be worked on with the covers removed must be equipped with an earth current trip.

Electrical cables and connectors

A common cause of electric shock and electrical fires is incorrectly rated or defective cable.

The electrical cables feeding appliances or equipment must be capable of handling the electric current demanded. Putting it simply, if the conductor is (a) not large enough, and (b) not made from the correct conductive material, it will heat up. The heating effect may start a fire directly, or it might melt the insulating material surrounding it causing a short circuit that might indirectly start a fire, or create an electric shock hazard. The factors governing the current handling capacity of an electrical cable will be discussed in more detail in Chapter 5.

Safety hazards exist where the outer protective sheathing is damaged causing the inner conductors to be exposed. Electric shock will result when a person either touches the conductors, or makes indirect contact should the conductors come into contact with any other metalwork or moist conditions.

The majority of domestic and office equipment is connected to the 230 V supply via a flexible two core or three core cable. When replacing a damaged cable it is essential that the correct type is used. There are primarily two points to consider: the cable rating, and the type of insulation. Cable rating is usually indicated in amps, for example 3 amp, 6 amp, 15 amp. The current rating of the existing cable should be ascertained, and an equivalent replacement selected.

PVC insulated mains cable is suitable for most applications where the equipment will be used indoors in a normal environment. Some PVC cables have flame retardant properties. Rubber insulation offers greater durability and is more flexible which makes it useful where long lengths are required. Heat resisting cables having textile or special rubber insulation are employed in situations where there is a chance that the cable might come into contact with a hot surface or implement, for example electric irons, soldering iron, electric heaters, etc.

Other factors such as insulation resistance are affected by the type of insulation used in a cable, and although outside the area of this book, it has been mentioned to further emphasise the importance of selecting the correct type of replacement cable.

The colour code for mains flexible connectors in the UK is **Brown** for **Live**, **Blue** for **Neutral**, and **Yellow/Green** for **Earth**. When working on equipment that was manufactured prior to the early 1970s, you might encounter cable using the old colour code which was Red for Live, Black for Neutral, and Green for Earth. During the 1970s there was a considerable amount of imported Japanese equipment that used a two core cable which was coloured Black for Live, and Blue for Neutral. When encountering these cables they should be replaced with an appropriate cable using the modern colour code.

When servicing electrical equipment, the mains cable should be inspected for any physical damage. This includes at the points where the cable enters the mains plug, and the equipment. If there are any signs of breaks in the insulation or broken entry clamps, the cable and/or the clamps should be replaced. The cable should also be checked for non-compliant extensions. This is where it has been extended using terminal block and insulation tape; or perhaps even more unorthodox methods. Because there is no clamp securing the two cables at the joint, such extensions do not comply with safety regulations because there is a chance of the joint being broken if pulled, and live wires being exposed. It is the responsibility of the service engineer to make the customer aware of the dangers of any such extensions and to recommend a suitable alternative, which in the majority of cases is to employ an approved mains extension cable.

Connection of equipment to the 230 V mains supply is usually made using an insulated plug, and for many years the three pin 13 amp mains plug has been the standard connector for most domestic equipment and appliances. This plug, illustrated in Figure 1.9, was introduced as a standard for domestic and commercial appliances to overcome the problems associated with different buildings having a multitude of different power sockets, and it has served its purpose well over the past thirty years. To comply with the British Standard BS 1363, the plug must be capable of handling up to 13 amps without overheating. The live connection must be fused with a cartridge fuse of a maximum rating of 13 amps; however, if the equipment to which it is connected operates at low power levels, a lower rated fuse must be fitted. Fuse selection will be discussed in more detail in Chapter 11.

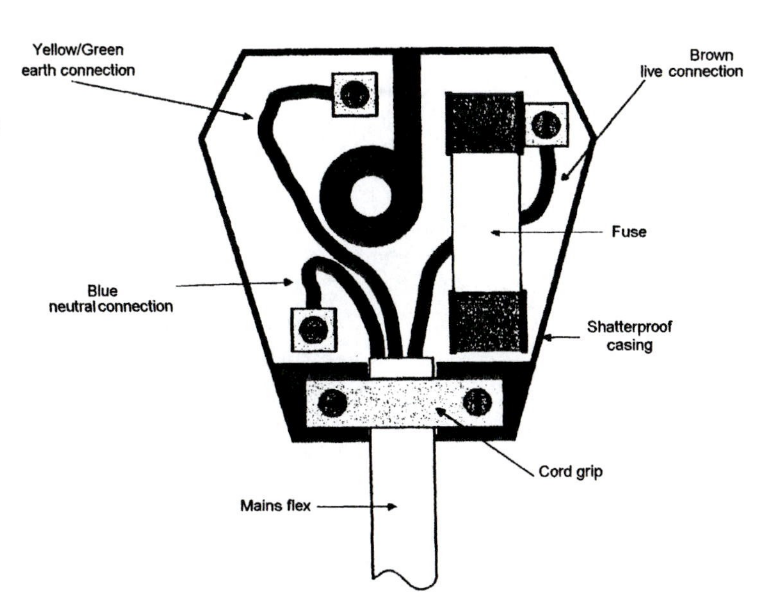

Figure 1.9 Fused 3 pin, 13 amp mains connector

Other standard features of this type of plug are the cord grip which is designed to secure the cable into the plug, shatterproof construction, and the extended earth pin which not only serves to open the safety shutters on the power socket, but also ensures that the earth connection is the first to be made, and the last to be broken.

Although the concept of the 13 amp mains plug was good, the main drawback proved to be the fact that the customer was expected to fit it. It has been said that any fool can fit a plug, the problem is that over the years many did, sometimes with disastrous consequences. More recently it became a requirement for all new equipment to come with a moulded plug fitted, thus reducing the number of occasions when an unqualified person has to perform the task.

The correct wiring configuration for the 13 amp plug is shown in Figure 1.9. When fitting these plugs, the engineer must check not only that the connections are the correct polarity, but also that there is no damage to the insulation of the cores or outer sheathing, no stray wires within the plug, the wires are dressed to fit and that none are trapped when the top is fitted, the wire lengths are such that if torn out the live will be the first to sever and the earth will be the last, the cable grip is secure, and the fuse rating is correct for the appliance to which it is connected.

The 13 amp plug serves to connect equipment to the wall socket; however, many items of equipment require a detachable mains lead, introducing the need for a suitable connector at the equipment end. The IEC 320 Standard 6 amp and 10 amp connectors are used for this purpose. As shown in Figure 1.10, a three pin moulded plug is fitted to the mains cable, although the live connections are actually embedded in sockets to provide insulation. The socket in the equipment contains three pins which make connection with the live plug.

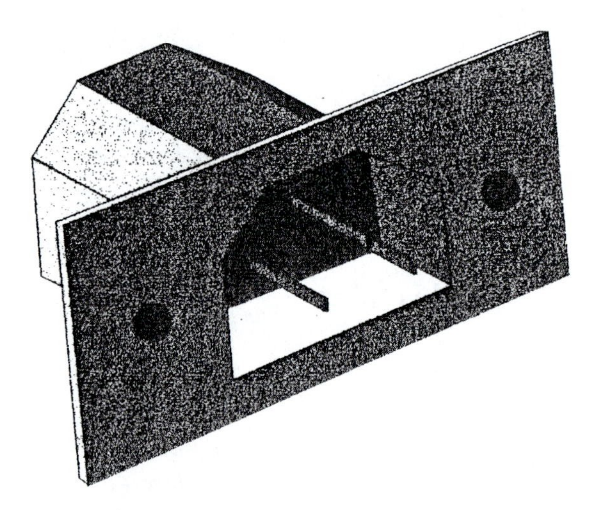

Figure 1.10

IEC chassis mount socket. Available in 6 A and 10 A ratings

Where a user has insufficient mains sockets, the problem can be overcome by using approved two- or three-way adapters, or multiple outlet distribution units. However, if used incorrectly these can present a safety hazard. Multiple outlet units are perfectly safe as long as the total current load does not exceed 13 A. So, for example, if a four-way mains block is used to connect a computer system comprising a base unit, a monitor, a printer, and a small sound system, there is no danger of the block being overloaded because the four items together will probably not draw more than 5 A. On the other hand, if the user plans to run two 1 kW electric heaters, a 3 kW kettle, and a 1 kW iron from such a block, the total current would be in the order of 30 A. This should cause the 13 A fuse in the plug connecting the block to the mains supply to blow, presenting no safety hazard. However, it is all too common practice for inexperienced DIY enthusiasts to replace the fuse with silver foil or some similar material which will quite happily pass the 30 A demanded by the load equipment. At this point a major fire hazard exists.

It is the responsibility of a service engineer to point out any such hazards to a customer. When required to install a system in a location where there are insufficient mains sockets, the engineer must calculate the total current load, and consider the most appropriate safe method of connection. In some cases this might involve calling in an electrician to add power sockets to the circuit.

The current load can usually be worked out by looking for the current rating of each piece of equipment, which should be printed on the outside of the casing. Where the rating is given in watts (power), a quick rule of thumb which errs on the side of caution is to say that 1 kW = 5 A.

Dealing with electrical shock

In the event of having to deal with someone who is suffering, or has suffered, an electric shock, the first thing that must be done is establish whether the person is still in contact with the electrical supply. If this is the case, then under no circumstances must anyone come into contact with that person, otherwise they too will more than likely suffer a shock.

The electrical supply must be isolated in some way. The best thing is to turn off the supply, perhaps by unplugging the equipment with which they are in contact, by operating a circuit breaker, or if these are not viable options, by wrenching the cable from the supply point.

If it is not possible to isolate the supply in any way, then attempt to push the person away from the supply using an insulated implement such as a wooden brush, a wooden chair, etc. Ideally you should be standing on an insulated surface such as wood, rubber, or plastic.

Once isolated, if the casualty is unconscious but breathing, loosen the clothing around the upper body and lay them in the recovery position. Check, and keep on checking, the breathing and pulse rates. As soon as possible send for help and ensure that an ambulance is called. Keep the casualty warm and comfortable.

In the event that the casualty's heart has stopped beating, begin cardiac compression.

If the casualty has suffered burns, and if it is practical to do so, immerse the burn into or under cold water until medical help arrives.

Fire safety

The best means of dealing with fires is to prevent them from occurring in the first place. Some common hazards in the field of electronics servicing are refilling gas charged soldering irons, using certain flammable aerosols such as lacquer and some contact cleaning agents, overloading electrical circuits, and leaving equipment on test unsupervised.

General points on fire safety are:

(i) Do not have naked flames or cigarettes in the workplace, especially when using flammable substances.
(ii) Ensure that all electrical supplies in the workplace are adequate, are not being overloaded, and are suitably maintained. Alert customers of any potential dangers, even if they are not directly related to the equipment you are concerned with.
(iii) The workplace should be adequately ventilated to prevent a build-up of hazardous fumes.
(iv) Suitable extinguishers, fire blankets and sand buckets should be available and adequately maintained. These must be clearly marked, and all employees must be aware of their location and use.
(v) Escape routes must be clearly marked. They must be kept clear at all times, and fire doors must be functional; that is, they must be able to open, and smoke doors must close and seal correctly.
(vi) Fire drills must be held regularly, and all fire fighting equipment must be inspected regularly.

Fire requires three ingredients: fuel, oxygen, and heat. Therefore when fighting a fire it is necessary to remove any one of these. Removal of the fuel will cause the fire to quickly burn itself out; however, quite often this is not the easiest option because the heat from the fire prevents you from getting close enough to remove the fuel. Removal of the oxygen supply will stop the fire. This is done by smothering the fire with a non-combustible or wet blanket, or with foam, carbon dioxide, sand, or other dry powder. Smothering might also be possible if the fire can be contained within a sealed area where the air supply is cut off. Removing the heat is done by cooling the fuel with water.

Figure 1.11 illustrates four types of fire extinguisher, along with the colour code and their main applications. It is important to use the correct type, otherwise hazards of an even greater magnitude than the actual fire may result. For example, using a water or foam extinguisher on an electrical fire where the circuit has not been isolated could result in electrocution. Using these extinguishers on a burning liquid fire usually causes the liquid to spread, carrying the flames with it. Conversely, a dry powder extinguisher is not very effective on a large paper or wood burning fire; water or CO_2 is far more effective.

Figure 1.11 Fire extinguishers and their applications

Type	*Colour*	*Applications*	*Notes*
Water	Red	Wood Paper Cloth	Class A fires. Direct at base of fire.
 Foam	Cream	Liquid	Class B fires. Direct above liquid so that the foam descends over it, smothering the fire.
Dry powder	Blue	Liquid Liquefied gas Electrical Burning clothing	Liquefied gas fires are best left to be tackled by trained personnel.
 Carbon dioxide (CO_2)	Black	Liquid electrical	

Upon discovering a fire, your first priority must be to raise the alarm and alert the fire brigade. If you are able to tackle the fire with the equipment available then you should do so, otherwise evacuate the premises along with all other personnel. **At no time should you put yourself or any other person at risk**. Make sure that if you do tackle the fire, you do so with the correct extinguisher or other fire equipment.

Lifting

One of the most common industrial injuries is back injury, and recovery is frequently a long process. Even where recovery appears complete, all too often a person can suffer from related problems later in life.

In electronics servicing it is frequently necessary to lift and transport heavy items, and service engineers need to be aware of their limitations in lifting. As a general rule, never attempt to lift equipment that is clearly too heavy, or simply too cumbersome for one person. Assess each situation as it arises and consider the options. Does the task require more than one person? Should the object be moved using mechanical equipment? Does it need to be moved at all?

The recommended method of lifting for a person working alone is known as the kinetic method. The posture for kinetic lifting is to place the feet 18" apart, with the knees slightly bent, keep the back straight and arms as close in to the body as possible. Lifting is then performed using the muscles in the legs, keeping the head erect.

The kinetic method is sometimes criticised because many items are the wrong shape to be lifted in this manner. However, there is no safer alternative, and the argument stands that if the object cannot be lifted in this manner then assistance should be sought. Unfortunately assistance, either in the form of another person or mechanical equipment, is all too often considered to be an unnecessary expense in either money and/or time. Yet it is a small price to pay in comparison to a lifetime of pain and discomfort.

2

The electronics servicing industry

It is difficult to fix a definite date for the birth of the electronics industry; however, it began to pick up momentum in the years following the First World War. As is so often the case, the war greatly accelerated technological development, bringing about quantum leaps in the development of thermionic valves, as well as radio and telecommunications technology. In the years following the war, research continued both in Europe and the USA into these relatively new fields.

Engineers and politicians alike became increasingly excited at the possibilities which were opening up in the fields of telecommunications and broadcast radio. Efficient international communications systems would revolutionise industry, and governments were keen to back development, no doubt urged by the prospect of an increased gross national product which they hoped would result from increased productivity, especially for the countries who were first to introduce the new technologies.

If governments were motivated by the possibilities of financial reward, engineers were spurred on by the wonders of the new technologies opening up before them. Suddenly the dreams of philosophers and science fiction writers were becoming reality. Forms of communication altered rapidly, having a profound effect on military, industrial and domestic life.

If the First World War gave a kick start to an already developing electronics industry, the Second World War launched it into the forefront of technology. Every leading nation involved in the war brought its electronics experts to work on development of communications and radar, as well as a host of other projects including the first electronic computer (made from valves, of course!).

Post war once again saw much of the newly discovered technology put back into civilian communications, industry, and domestic leisure. The world was becoming a safer place because of new communications systems. Industry was becoming more automated. Domestic life changed dramatically as television brought the world into the front room of everyone.

Yet more significantly, the rate at which this relatively new industry was developing continued to increase. Semiconductor technology was slow to be accepted by many to begin with; however, within two decades it was bringing sweeping changes into the world of electronics. Furthermore, the space race of the 1960s and 1970s diverted billions of American dollars into research of electronic component miniaturisation, resulting in spin-offs that have changed the way the western world lives and works. Microprocessor chips, fibre optic links, laser technology, and of course satellite communications have all had a profound effect on human society.

Servicing

As the manufacturing industry grew in the post First World War years, so another industry was born and began to develop. Servicing.

To begin with service engineers were often self-taught and worked alone. But as technology leapt forward and the amount of electronic equipment in the western world increased, the need for a more organised army of suitably trained service engineers became apparent. Organisations geared specifically to the servicing of electronic equipment spawned, encouraged to some extent by the manufacturing industry which saw the need for service back-up for its products.

Traditionally an electronics service engineer was seen as a blue-collar worker armed with a toolbox, a test meter, and a soldering iron. Whilst still requiring all of these items, the modern engineer may now have access to data banks containing service information relating to the product he is working on. Moreover, this information can be accessible at the place of work, and not just back in the workshop. An increasing number of electronic adjustments are now carried out using a PC and a suitable interface, the PC program offering useful information to the service engineer. Fast courier services mean that spare parts can frequently be ordered from the manufacturer and dispatched directly to the customer's premises, arriving often within 24 hours. All of these services make the laptop computer and mobile phone as much a part of an engineer's kit as screwdriver and side cutters.

There are those who voice the opinion that electronic servicing as an industry is in its dying throes; that because of increased reliability of equipment and greatly reduced production costs the need for service engineers will continue to reduce. Certainly the nature of servicing has altered drastically over the years with modular replacement becoming more the norm, and there is no doubt that increased reliability has cut the number of people employed in the service industry. However, there are still many thousands of people earning a living maintaining and repairing all types of electronic and electromechanical equipment. Equipment without which modern offices and industries would be unable to function.

Electronics servicing plays a vital role in both industry and society in any modern industrialised nation.

Manufacturing

The manufacture of electronic equipment is considerably different to servicing; however, a service engineer can benefit from understanding the manufacturing processes, and a visit to a modern production plant can be very illuminating.

Even manufacture on a medium scale requires a lot of space, manpower, and organisation, not to mention a considerable amount of capital investment. Because of media interest in fully automated production lines it is easy to have the impression that *all* modern production lines are like this. While this is generally true for very large-scale production runs, the high cost of purchase and setting up of an automated line makes its cost prohibitive for equipment that is to be produced in smaller numbers. A considerable amount of electronic equipment is produced using a hybrid of new and older style production techniques. For example, automated methods are generally used for the production of printed circuit boards, small component insertion onto the boards, and final soldering of the components. However, medium and large size components are frequently inserted and soldered by hand, and final assembly is often much more cost effective when done manually.

The operation of a manufacturing company is complex. A traditional company structure is illustrated in Figure 2.1.

Figure 2.1
Traditional manufacturing company structure

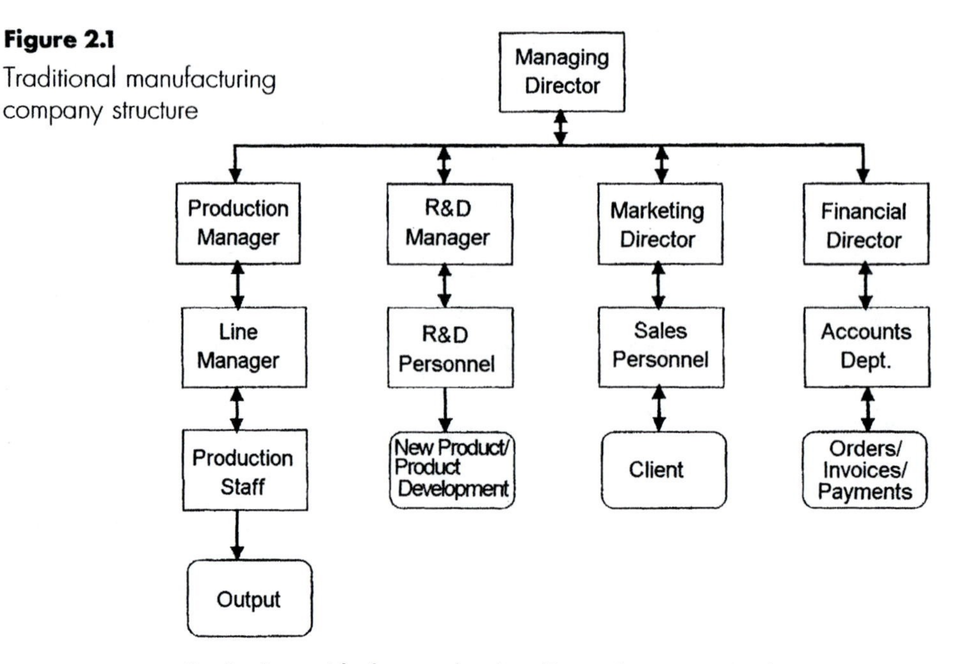

Beginning with the production lines, these require line managers to monitor the process. They must investigate any problems that occur either in the production process, or with the actual equipment being produced. Problems that cannot be resolved quickly are reported to the production manager who can liaise with other departmental managers. The production manager is also responsible for stock control, that is, ensuring that there are sufficient stocks of component parts to maintain production, and quality control.

Of course, there is no point in producing something and not telling anyone that it exists! Every company needs a good marketing team, which usually comprises a marketing director who heads a group of sales personnel. Marketing is most successful when there is clear two-way communication between the marketing team and another important group of people, the research and development (R&D) team.

Electronic devices, equipment, and supporting software are evolving at an incredible rate, and every company needs to keep at the forefront of technology if they intend to stay in business. The R&D team have the challenging task of developing new products, and continually improving or evolving existing ones. The R&D manager must communicate regularly with the production manager. Problems on the production line are frequently only remedied back in the laboratory. On the other hand, the production manager can often help avert problems by examining products under development and offering advice on how best to design something that can actually be produced. It has often been said that the design technician produces the design, but it is the engineer who makes it work!

The success of many companies can be attributed to clear two-way communication between the marketing staff and the R&D people. If the sales people are familiar with the technical aspects of the product they are selling, then they are able to talk far more intelligibly to their clients. On the other hand, clients' views must be fed back to the R&D team in order that they can develop something that people actually require, rather than what they may think they require.

The financial director has the task of ensuring that the company's finances are managed such that it can continue to function and stay solvent. Because all departments need money to function, all managers must communicate with the financial director. Perhaps the most significant communication is between the financial and marketing directors.

At the top of this hierarchy is the managing director (MD) whose job it is to monitor the work of the other directors. Responsible ultimately to the shareholders, the MD will meet with the other directors regularly to monitor and discuss the performance and progress of the company. It is the board of directors who consider new ideas, which may originate either from the R&D section or from outside the company, and decide which should be pursued.

And finally, there are the many other support staff including clerical, catering, warehouse, cleaning, security, drivers, and many others without whom a company would falter.

The traditional model shown in Figure 2.1 cannot be adhered to in small or even many medium size companies. Such a high staffing level in the management structure would be unaffordable. Never the less, for a company to function efficiently the individual roles outlined in this structure are necessary, and where a smaller company is concerned it is often possible for one person to perform more than one role. For example, all marketing may be done by one person, who may also have an involvement with the company accounts.

Even larger manufacturing companies in recent years have had to move away from the traditional model due to financial constraints which have resulted in reductions in staffing levels.

3

Calculations

The ability to perform some simple calculations is a necessity for any electronics engineer. It can be viewed as just another useful tool which makes work easier, hence a little time spent in grasping some fundamental basics will not be wasted. The following information is intended to serve as a reference with information of how to perform a variety of calculations, but the methods demonstrated are only examples since there are many different methods that can be used to do the same calculation.

Fractions

A quarter written as $\frac{1}{4}$ is a fraction. If a cake is taken and divided equally into four pieces, and one of these pieces is given to you, you will have one fourth of the cake. One fourth is called a quarter which is written as $\frac{1}{4}$. There are four quarters $\frac{4}{4}$ in the cake or, if referring to numbers only, the whole which equals 1. How many fiftieths are there in one whole or 1? The answer is fifty. From the examples used so far it can be seen that the lower of the two numbers of a fraction describes how many parts the whole has been divided into:

$\frac{1}{50}$ means that there are 50 equal parts

$\frac{1}{75}$ means that there are 75 equal parts

$\frac{1}{100}$ means that there are 100 equal parts

The lower number (that describes the number of equal parts) is called the denominator while the upper of the two numbers (the one which states the number of parts given to us) is called the numerator. For example $\frac{3}{100}$ means that the whole has been divided into 100 equal parts and we have 3 of them.

Adding fractions

Suppose that you have been given one piece of a cake that has been divided equally into four parts and another piece from a second cake that has been cut into 16 parts. How much cake do you have?

$$\frac{1}{4}+\frac{1}{16}=$$

It is not possible to add together two fractions which have different denominators. The denominator is not a value but an indicator of how many parts a whole has been divided into. To carry out this addition it is necessary to find a common denominator:

$\frac{1}{16}$ is 1 part of 16

$\frac{1}{4}$ is 1 part of 4

If the smallest denominator will divide exactly into the largest denominator then the larger of the two denominators is the common denominator since this is the smallest number that both will divide into. Four will divide into sixteen, four times. This tells us that $\frac{1}{4}$ is four times larger than $\frac{1}{16}$; alternatively it can be suggested that to change 4 into 16 it was multiplied by four. To keep its true value a fraction must have both its denominator and its numerator increased or decreased in the same way and by the same factor:

$\frac{1}{4}=\frac{\quad}{4\times 4}=\frac{\quad}{16}$ Denominator × 4

$\frac{1}{4}=\frac{1\times 4}{\quad}=\frac{4}{\quad}$ Numerator × 4

$$\frac{1}{4}=\frac{1\times 4}{4\times 4}=\frac{4}{16}$$

The common denominator of $\frac{1}{4}$ and $\frac{1}{16}$ is 16. It is not the only common denominator. Other examples could be 32, 48, 64. Sixteen is the smallest number that both 4 and 16 will divide into and is called the Lowest Common Denominator – LCD.

Find the LCDs of the following (answers at the end of the chapter):

1. $\frac{1}{5}$ and $\frac{1}{15} =$

2. $\frac{1}{6}$ and $\frac{1}{12} =$

3. $\frac{1}{7}$ and $\frac{1}{21} =$

4. $\frac{1}{1.8}$ and $\frac{1}{54} =$

5. $\frac{1}{40}$ and $\frac{1}{400} =$

In all the cases the smaller denominator will divide directly and equally into the larger. Consider the following fractions, $\frac{1}{6}$ and $\frac{1}{9}$. The smallest number that both 6 and 9 will divide equally into is 18. How have we achieved this? Basically from remembered experience, but without experience this is not possible. Since 6 will not divide equally into 9 an alternative is to multiply the two denominators together:

$6 \times 9 = 54$

Both numbers will now divide exactly into 54 and our addition can continue but some extra work may be required later on. This will be to get the fraction in its lowest terms, but this will be dealt with later. Returning to the problem of adding two fractions, it is necessary to ensure the denominators are the same.

Examples

1. $\frac{1}{4} + \frac{3}{8}$

Each fraction has to have the same denominator. Since 4 will divide into 8 the common denominator can be 8. We now have to change a quarter into an equivalent value with 8 as its denominator. To change 4 into 8 it has been multiplied by 2. The numerator 1 must be multiplied by 2.

Hence $\frac{1}{4} = \frac{1 \times 2}{4 \times 2} = \frac{2}{8}$

It is now possible to add $\frac{1}{4}$ to $\frac{3}{8}$

$$\frac{1}{4} + \frac{3}{8} = \frac{2}{8} + \frac{3}{8} = \frac{5}{8}$$

2. $\frac{1}{3} + \frac{1}{4}$

The smallest denominator (3) will not divide into the larger (4). The common denominator to use is therefore the smaller multiplied by the larger – $3 \times 4 = 12$

To change one third into an equivalent value with twelve as its denominator, both the denominator and numerator must be multiplied by 4:

$$\frac{1}{3} = \frac{1 \times 4}{3 \times 4} = \frac{4}{12}$$

To change $\frac{1}{4}$ into twelfths multiply both the numerator and denominator by 3:

$$\frac{1}{4} = \frac{1 \times 3}{4 \times 3} = \frac{4}{12}$$

Therefore $\frac{1}{3} + \frac{1}{4} = \frac{4}{12} + \frac{3}{12} = \frac{7}{12}$

So far only two fractions have been added together. How then can more than two fractions be added?

Example

$$\frac{1}{4} + \frac{1}{3} + \frac{1}{5}$$

Find the common denominator:

$4 \times 3 \times 5 = 60$

$\frac{1}{4} = \frac{15}{60}$ To change 4 into 60 it must be multiplied by 15.

Remember, if the denominator is multiplied by 15 so too must the numerator:

$\frac{1}{3} = \frac{}{60}$ To change 3 into 60 it must be multiplied by 20.

Hence $\frac{1 \times 20}{3 \times 20} = \frac{20}{60}$

Therefore $\frac{1}{5} = \frac{12}{60}$

All fractions now have the same denominator so the addition can continue:

$$\frac{1}{4} + \frac{1}{3} + \frac{1}{5} = \frac{15}{60} + \frac{20}{60} + \frac{12}{60} = \frac{47}{60}$$

OR $\qquad = \frac{15 + 20 + 12}{60} = \frac{47}{60}$

Add the following (answers at the end of the chapter):

6. $\frac{1}{3} + \frac{3}{8} =$

7. $\frac{1}{6} + \frac{2}{16} + \frac{3}{7} =$

8. $\frac{1}{5} + \frac{3}{15} + \frac{7}{60} =$

9. $\frac{6}{11} + \frac{1}{5} + \frac{2}{55} =$

10. $\frac{7}{8} + \frac{2}{16} + \frac{5}{32} =$

Is there anything unusual about the answer to example 10 above? There should be. Your answer to number 10 is called an *improper fraction.*

An improper fraction is a fraction where the numerator is bigger than the denominator. These improper fractions can be changed into a whole number plus a proper fraction. It is then called a *mixed number.* For example, to change $\frac{15}{8}$ into a mixed number, the denominator is divided into the numerator (bottom number into top number):

$15 \div 8 = 1$ remainder 7

The remainder (7) is put over the denominator to give $\frac{7}{8}$:

$\frac{15}{8}$ is therefore equal to $1\frac{7}{8}$

Now do the same for your answer to example 10 above.

Further examples

Change the following to mixed numbers (answers at the end of the chapter):

11. $\frac{4}{3} =$

12. $\frac{10}{3} =$

13. $\frac{17}{2} =$

Adding mixed numbers

Add the following mixed numbers $1\frac{6}{7} + 2\frac{1}{14}$, add the whole numbers first: 1 + 2 = 3. Now add the fractions using the rules already learnt:

$$\frac{6}{7} + \frac{1}{14} = \frac{12}{14} + \frac{1}{14} = \frac{13}{14}$$

So $1\frac{6}{7} + 2\frac{1}{14} = 3\frac{13}{14}$

Add the following (answers at the end of the chapter):

14. $1\frac{5}{8} + 2\frac{5}{16}$

15. $3\frac{4}{5} + 7\frac{5}{8}$

16. $6\frac{7}{8} + 7\frac{1}{4}$

17. $9\frac{3}{4}+\frac{2}{3}$

18. $7\frac{1}{2}+1\frac{6}{7}$

Subtraction

As with addition it is not possible to subtract fractions with different denominators. The first step therefore is to determine the common denominator and then subtract the numerators.

For example, $\frac{1}{2}-\frac{1}{4}$ Common denominator = 4

$=\frac{2-1}{4}=\frac{1}{4}$

19. $\frac{2}{3}-\frac{1}{4}=$

20. $\frac{6}{7}-\frac{5}{14}=$

21. $\frac{7}{9}-\frac{2}{18}=$

22. $\frac{19}{23}-\frac{2}{69}=$

23. $\frac{5}{6}+\frac{1}{24}=$

(answers at the end of the chapter)

Mixed numbers

$2\frac{1}{2}-1\frac{1}{8}$

Just as when carrying out an addition deal with the whole numbers first: $2-1=1$

Now deal with the fraction:

$\frac{1}{2}-\frac{1}{8}=\frac{4-1}{8}=\frac{3}{8}$

Answer = $1\frac{3}{8}$

There may be an added difficulty. Consider the following example: $3\frac{1}{4} - 1\frac{5}{8}$. Dealing with the whole numbers first produces the following: $3 - 1 = 2$. Now dealing with the fractions:

$$\frac{1}{4} - \frac{5}{8} = \frac{2-5}{8}$$

5 cannot be subtracted from 2. One of the whole numbers is now changed to eighths: $2 = 1 + \frac{8}{8}$. The whole number part of the answer is now 1. The fractional part now becomes:

$$\frac{8+2-5}{8} = \frac{10-5}{8} = \frac{5}{8}$$

The answer is therefore $1\frac{5}{8}$.

We can now *add* and *subtract* fractions.

Test (answers at the end of the chapter):

24. $\frac{7}{8} - \frac{5}{16} =$

25. $2\frac{1}{2} + \frac{3}{16} =$

26. $1\frac{3}{4} + \frac{1}{3} =$

27. $2\frac{1}{3} + 1\frac{5}{8} =$

28. $3 - 4\frac{1}{8} + 2\frac{1}{3} =$

29. $2\frac{1}{4} - 3\frac{1}{2} + 2\frac{3}{4} =$

30. $1\frac{9}{16} - 2\frac{1}{2} + \frac{1}{8} =$

31. $2\frac{11}{16} - 1\frac{1}{4} =$

32. $\frac{11}{16} - 2\frac{1}{4} + 3\frac{5}{16} =$

33. $2\frac{1}{2} - 2\frac{1}{4} + \frac{11}{16} + \frac{7}{8} =$

Multiplication of fractions

There are two cases to consider:

1. *Proper fraction*–A fraction where the numerator is smaller than the denominator, i.e. $\frac{3}{8}$.
2. *Mixed numbers*–Where there is a whole number and a fraction, $1\frac{5}{8}$.

Proper fractions

If two fractions are to be multiplied together, e.g. $\frac{1}{50} \times \frac{1}{10}$, the numerators are multiplied together (1×1), and the denominators are multiplied together (50×10):

$$\frac{1}{50} \times \frac{1}{10} = \frac{1 \times 1}{50 \times 10} = \frac{1}{500}$$

Example:

Multiply $\frac{1}{50} \times \frac{3}{8} = \frac{1 \times 3}{50 \times 8} = \frac{3}{400}$

Work through the following example: $\frac{2}{9} \times \frac{3}{7} = \frac{6}{63}$.

Consider the following example: $\frac{2}{3} \times \frac{9}{30}$. Using the method already demonstrated the result would be:

$$\frac{2}{3} \times \frac{9}{30} = \frac{2 \times 9}{3 \times 30} = \frac{18}{90}$$

However, $\frac{18}{90}$ is not in its simplest form since 18 and 90 can both be divided by the *same* number. It might not be immediately obvious that 18 will divide into itself once, and into 90 five times. The answer should reflect this and be in its simplest form:

$$\frac{18}{90} = \frac{18 \div 18}{90 \div 18} = \frac{1}{5} \rightarrow \text{Answer A}$$

Alternatively, since both 18 and 90 are even numbers they can be divided by 2,

$18 \div 2 = 9$, $90 \div 2 = 45$, hence $\frac{18}{90} = \frac{9}{45}$. Remember dividing the top and bottom numbers of a fraction by the same amount does *not* change its value.

Can $\frac{9}{45}$ be simplified further–Yes.

Again, it might not be obvious to you that 9 will divide into itself once and into 45 five times.

$9 \div 9 = 1 \rightarrow 45 \div 9 = 5$

Therefore $\frac{9}{45} = \frac{1}{5}$ The same answer as A.

Again, it is not obvious that 9 is a common divisor, perhaps a smaller divisor should be tried. Both 9 and 45 are odd numbers and therefore will *not* divide by 2. Try the next smallest number 3:

$\frac{9}{45}$ $9 \div 3 = 3$ $45 \div 3 = 15$

So that $\frac{9}{45} = \frac{3}{15}$ → Answer B

Again, if you have obtained Answer B it is necessary to decide if it is in its simplest form. Both 3 and 15 are odd numbers so dividing by 2 will not work. Try the next smallest number 3.

$3 \div 3 = 1$ $15 \div 3 = 5$

the answer is therefore $\frac{1}{5}$ the same as Answer A.

All the answers obtained are correct since

$$\frac{2}{3} \times \frac{9}{30} = \frac{18}{90} = \frac{9}{45} = \frac{3}{15} = \frac{1}{5}$$

$\frac{1}{5}$ is the best answer since it is in the simplest form. However, don't worry if you can't get from $\frac{18}{90}$ to $\frac{1}{5}$ in one step, as we have demonstrated dividing by common smaller numbers will still enable you to achieve the simplest answer. Multiply the following and give the answers in their simplest form:

34. $\frac{3}{4} \times \frac{4}{9} =$

35. $\frac{3}{4} \times \frac{2}{3} =$

Multiplication can also be achieved through a process called *cancelling*. Consider the same example:

$$\frac{2}{3} \times \frac{9}{30}$$

The 3 can be divided into itself once. The 9 will also divide by 3 to give an answer of 3. For example, this is normally shown as below:

$$\frac{2}{\cancel{3}_1} \times \frac{\cancel{9}^3}{30}$$

The example now looks:

$$\frac{2}{1} \times \frac{3}{30}$$

Further cancellations are possible since 3 will divide into both 3 and 30. It now appears as follows:

$$\frac{2}{1} \times \frac{\not{3}^{1}}{\not{30}_{10}} = \frac{2}{1} \times \frac{1}{10}$$

Are there any further cancellations – yes, 2 will divide into 2 and 10:

$$\frac{\not{2}^{1}}{1} \times \frac{1}{\not{10}_{5}} = \frac{1}{1} \times \frac{1}{5} = \frac{1}{5}$$

No further cancelling is possible so the calculation is finished. It is not necessary to rewrite the calculation at every step, all cancelling can be shown at one step.

For example,

$$\frac{\not{2}^{1}}{\not{3}_{1}} \times \frac{\not{9}^{3}}{\not{10}_{5}} = \frac{1 \times 3}{1 \times 5} = \frac{3}{5}$$

3 is cancelled in 3 and 9 while 2 is cancelled in 2 and 10.

Multiply the following (answers at the end of the chapter):

36. $\frac{2}{7} \times \frac{6}{11} =$

37. $\frac{3}{5} \times \frac{2}{7} =$

38. $\frac{7}{9} \times \frac{2}{4} =$

39. $\frac{5}{8} \times \frac{4}{15} =$

40. $\frac{3}{4} \times \frac{8}{9} =$

Mixed numbers

A mixed number is one which contains a *whole* number and a *fraction*, i.e. $2\frac{1}{2}$. To carry out a multiplication which contains a mixed number it is essential to convert the mixed number into an improper fraction. Convert $2\frac{1}{2}$ into an improper fraction $2\frac{1}{2} = \frac{5}{2}$. This was achieved by multiplying the whole number (2) by the denominator (2) and then adding the numerator of the fraction to the result. This is then placed over the denominator of the fractional part, i.e. $2 \times \frac{}{2}$(denominator) $= 4$, $4 + \frac{1}{}$(numerator) $= 5$. Place the result over the original denominator (2), hence $2\frac{1}{2} = \frac{5}{2}$. It is now possible to perform a multiplication using this number.

Example

$2\frac{1}{2} \times \frac{1}{3}$ Change $2\frac{1}{2}$ to an improper fraction

$$2\frac{1}{2} = \frac{5}{2} \times \frac{1}{3} = \frac{5 \times 1}{2 \times 3} = \frac{5}{6}$$

All the standard rules of multiplication can be applied.

Multiply the following (answers at the end of the chapter):

41. $\frac{2}{3} \times \frac{15}{16} =$

42. $\frac{4}{5} \times \frac{6}{7} =$

43. $1\frac{1}{3} \times \frac{3}{5} =$

44. $3\frac{5}{6} \times 2\frac{1}{2} =$

45. $1\frac{5}{9} \times 4\frac{2}{7} =$

Consider the answers to numbers 44 and 45. After multiplication both these answers are improper fractions; remember, it is necessary to change them into mixed numbers.

Division

To divide one fraction by another, all that is required is to *invert* the divisor and then perform a multiplication. Note – a mixed number must be turned into an improper fraction. To invert $\frac{1}{2}$ all that is required is to turn the fraction upside down so that it becomes $\frac{2}{1}$.

Invert the following:

1. $\frac{6}{2} = \frac{2}{6}$

2. $\frac{2}{3} = \frac{3}{2}$

3. $\frac{5}{7} = \frac{7}{5}$

Divide the following:

1. $\frac{3}{4} \div \frac{1}{2}$ Invert $\frac{1}{2}$

$\frac{3}{\not{4}_2} \times \frac{\not{2}^1}{1}$ Use cancelling

$\frac{3}{2} \times \frac{1}{1}$ Multiply

$\frac{3 \times 1}{2 \times 1} = \frac{3}{2}$ Change to a mixed number

2. $2\frac{3}{4} \div 1\frac{1}{4}$ Change both mixed numbers to improper fractions

$2\frac{3}{4} = 2 \times 4 = 8 + 3 = 11$ Denominator 4 hence $\frac{11}{4}$

$1\frac{1}{4} = 1 \times 4 = 4 + 1 = 5$ Denominator 4 hence $\frac{5}{4}$

$2\frac{3}{4} \div 1\frac{1}{4} = \frac{11}{4} \div \frac{5}{4}$ Invert $\frac{5}{4}$: change $\div$ to $\times$

$= \frac{11}{\not{4}1} \times \frac{\not{4}^1}{5} = \frac{11}{5}$ Use cancelling and change to mixed number

$= 2\frac{1}{5}$

3. $2\frac{1}{4} \div \frac{2}{3}$ — Change mixed number to improper fraction

$2\frac{1}{4} = 2 \times 4 = 8 + 1 = 9$ — Denominator 4 hence $\frac{9}{4}$

$\frac{9}{4} \div \frac{2}{3}$ — Invert $\frac{2}{3}$; change $\div$ to $\times$

$\frac{9}{4} \times \frac{3}{2} = \frac{27}{8}$

$= 3\frac{3}{8}$ — Change mixed number to improper fraction

4. $10 \div 3\frac{1}{3}$

This is a new situation since 10 is a whole number and not a fraction or mixed number. However, 10 can be written as a fraction by placing it over the denominator 1, so that 10 becomes $\frac{10}{1}$. The problem now becomes: $\frac{10}{1} \div 3\frac{1}{3}$. Now complete the problem in the normal way:

$\frac{10}{1} \div 3\frac{1}{3}$

There is one final point to deal with and this is priorities.

Priorities

If the following calculation is to be performed in what order should the tasks be carried out: $(1\frac{2}{3} + \frac{3}{4}) \times (\frac{1}{2} + 1\frac{1}{4})$. The order to perform calculations must always be:

1. Complete the work in brackets.
2. Complete the multiplications and divisions.
3. Complete the additions and subtractions.

Examples

1. $1\frac{3}{4} \times \left(1\frac{1}{2} - \frac{1}{6}\right)$ — Complete the work in brackets

$= 1\frac{3}{4} \times \left(1\frac{3-1}{6}\right)$

$= 1\frac{3}{4} \times 1\frac{2}{6}$ — Note: $1\frac{2}{6} = 1\frac{1}{3}$

$= 1\frac{3}{4} \times 1\frac{1}{3}$ — Change to improper fractions

$$= \frac{7}{\not{4}_1} \times \frac{\not{4}^1}{3} = \frac{7 \times 1}{1 \times 3}$$

$$= \frac{7}{3} = 2\frac{1}{3}$$

2. $\frac{5}{16} \times \left(2\frac{1}{4} - 1\frac{3}{4}\right)$ Complete the work in brackets

$\frac{5}{16} \times \left(3\frac{1+3}{4}\right)$

$\frac{5}{16} \times 3\frac{4}{4} = \frac{5}{16} \times (3+1)$ Note that $\frac{4}{4} = 1$

$\frac{5}{16} \times 4 = \frac{5}{16} \times \frac{4}{1}$

$= \frac{20}{16} = 1\frac{4}{16} = 1\frac{1}{4}$

There is one other term to consider – *OF*

For example, $\frac{2}{3}$ OF $\frac{1}{2}$

To solve this simply replace the term OF by the multiplication sign × so that

$\frac{2}{3}$ OF $\frac{1}{2} = \frac{2}{3} \times \frac{1}{2} = \frac{2}{6} = \frac{1}{3}$

Examples (answers at the end of the chapter):

46. $1\frac{1}{4} \div \frac{1}{2} + \frac{1}{2}$

47. $2\frac{1}{2} - \left(1\frac{1}{2} \times 1\frac{1}{4}\right)$

48. $\left(\frac{2}{3} + \frac{3}{4}\right) \times 1\frac{1}{2} + \frac{1}{2}$

49. $\frac{5}{8}$ OF $80 - 16$

50. $\frac{3}{4}$ OF $6 - 2$

Decimals

Numbers that are used daily are in this form – money, measurements, etc. Their correct name would really be *decimal fractions*. However, the normal method of using a numerator and a denominator is not employed. Instead a *radix* point called the *decimal point* is included. Numbers that appear on the right-hand side of this decimal point are numerators whose denominators are 10, 100, 1000, etc.

Thousands Hundreds Tens Units • Tenths Hundredths

For example, $1064.125 = 1064\frac{125}{1000}$

It can be broken down into its various components:

1064	=	One Thousand	1000
		No Hundreds	000
		Six Tens	60
		Four Units	4
			1064
While 0.125	=	One Tenth	$\frac{1}{10}$
		Two Hundredths	$\frac{2}{100}$
		Five Thousands	$\frac{5}{1000}$

Using your knowledge of fractions add the above. The answer is $\frac{125}{1000}$.

Changing the following to decimals shows that:

1. $\frac{1}{10} = 0.1$
2. $\frac{4}{10} = 0.4$
3. $\frac{36}{100} = 0.36$
4. $\frac{95}{100} = 0.95$
5. $\frac{387}{1000} = 0.387$

If the value given is an improper fraction apply the standard rules learnt for fractions.

For example, $\frac{469}{100} = 4\frac{69}{100} = 4.69$

$$\frac{902}{100} = 9\frac{2}{100} = 9.02$$

Addition and subtraction

The method employed is the same as that used for whole numbers *but* it is *important* to keep the *decimal point in line.*

For example, 1.69−0.35

$$\begin{array}{r} 1.69 \\ -\underline{0.35} \\ \underline{1.34} \end{array}$$

It would, however, be most likely that this calculation would be performed using a calculator.

Multiplication

Again it has to be said that this operation is likely to be performed using a calculator but it is useful to appreciate how the answer is achieved by normal means.

For example, 2.4 × 4

Ignore the decimal point and multiply as whole numbers, i.e. 24 × 4 = 96. The decimal point can be inserted last, remember the original problem 2.4 × 4. How many figures were to the right of the decimal point – ONE. There must be one decimal place in the answer, 96 becomes 9.6, repeat this exercise for the problem 3.8 × 5.2, ignore the decimal points.

Multiply as whole numbers:

38 × 52	38	
	52	
	1900	multiplying 38 by 50
	+ 76	multiplying 38 by 2
	1976	addition of two products

Now reinsert the decimal point. How many figures were to the right of the decimal point?

3.8 – ONE   5.2 – ONE   TOTAL = TWO

The answer must contain *two* decimal figures, so that 1976 becomes 19.76.

Division

This is usually performed by the standard method known as *long division* or by using a calculator. Consider the following example done by long division, 76.8 ÷ 8; always write down the problem and place the decimal point in the answer, immediately above the one in the number being divided:

```
   .
8)76.8
```

Division can now begin. It is done a step at a time, dividing each digit in turn, starting on the left, 8)7. Since 8 will not divide into 7 the next digit (6) is placed alongside 8)76. 8 will divide into 76 nine times with some left over. To establish how much is left over a simple method is used:

```
  9.
8)76            multiply 8 by 9 and write below
```

```
  9.
8)76
 -72            subtract 72 from 76
   4
```

Note that the remainder (4) is less than the divisor (8). To continue with the calculation the next digit is brought down to sit alongside the 4:

```
   9.
8)76.8
  72
   48
```

Eight is now divided into 48. The answer is 6:

```
   9.6
8)76.8
 -72             (9 × 8 = 72)
   48
  -48            (6 × 8 = 48)
```

76.8 ÷ 8 therefore equals 9.6

Example

849.42÷27

```
     31.46      Insert decimal point, 8÷27
27)849.42       not possible
   -81          84 ÷ 27 = 3 × 27 = 81. Place 3 at top
    39          bring down next digit and ÷
   -27          1 × 27, place 1 at top, bring down next digit and ÷
    124         124 ÷ 27 = 44 × 27 = 108, place 4,
   -108         bring down next digit and:
     162        162 ÷ 27 = 6, place 6 at top.
    -162
     ...        No remainder, division complete.
```

Answer= 31.46

Not all examples are quite as simple, for example 75.66 ÷ 0.6. Before this can be performed the 0.6 must be changed into a whole number by moving the decimal point one place to the right, 0.6 = 6, this is the same as multiplying 0.6 by 10. To keep everything even the decimal point must be moved one place to the right in the other number so that 75.66 becomes 756.6, so 75.66 ÷ 0.6 now becomes 756.6 ÷ 6. Complete the division in the way demonstrated.

Further examples (answers at the end of the chapter):

51. 747 ÷ 0.9

52. 0.1118 ÷ 4.3

53. 181.7 ÷ 2.3

Corrected decimals

Suppose the answer to a problem is 265.1276431. The question has to be, how important are the last few digits (6431)? If the calculation was concerned with money then all digits beyond 265.12 would have very little real meaning. However, it is possible to take into account these values by *rounding off*. It is possible to round off to two decimal places and then say that the number quoted is correct to two decimal places. In rounding off a simple rule is applied, if the first figure to be disregarded is five or more then add one to the figure to its left:

27.666 = 27.67 correct to 2 decimal places

If the first figure to be disregarded is 4 or less the last figure stays the same:

27.664 = 27.66 correct to 2 decimal points

Correct the following to one decimal place (answers at the end of the chapter):

54. 7.68

55. 1.018

56. 28.97

57. 10.84

58. 8.048

59. 3.27

Indices

If two numbers are multiplied together, the answer produced is usually in it shortest form, i.e. $3 \times 2 = \underline{\underline{6}}$, $7 \times 3 = \underline{\underline{21}}$.

However, if a number is multiplied by itself, 5×5, the answer 25 can be written in a shorter form – 5^2 – five squared or *five to the power two.*

The figure above and to the right of the 5 is called an *index*. The index indicates how many times the number is multiplied by itself:

9^5 means 9 multiplied by itself five times

$9^5 = 9 \times 9 \times 9 \times 9 \times 9$

10^4 means 10 multiplied by itself four times

$10^4 = 10 \times 10 \times 10 \times 10$

10^4 is therefore a short way of writing 10 000

In the above example (10^4) the 10 is described as the *base* while the 4 is termed the *index* or *power*. This is a convenient shorthand form. It is easier to say or write 10^4 rather than *a number to the base 10 and a power of four* or *a whole number of 10 and a figure above and to the right of the whole number with a value of four.* The short form is also technically more correct. This will become more apparent later when voltages and currents are measured, i.e. 3 mA is the engineers way of writing 3 milliamperes.

Express $10 \times 10 \times 10 \times 10 \times 10$ as a base and index. Answer $= 10^5$

So 10^5 is a number with a base of 10 and a power of 5, 10^5 is the short way of writing 100 000.

$1\,000\,000 = 10 \times 10 \times 10 \times 10 \times 10 \times 10$

$= \underline{\underline{10^6}}$

Complete the following examples:

10 000 = 10 to the power

100 = 10 to the power

10 = 10 to the power

Can you see the connection between the number of noughts in relation to the index when the base is 10. *The index is always the same as the number of noughts.* NB – Only when the base is 10.

Consider the multiplication:

$10^2 \times 10^3$

$= 10 \times 10 \times 10 \times 10 \times 10$

$= 100\,000$ OR 10^5

Repeat this exercise to multiply 10^4 by 10^5:

$10^4 = 10 \times 10 \times 10 \times 10$

$10^5 = 10 \times 10 \times 10 \times 10 \times 10$

$10^4 \times 10^5 = 10 \times 10 \times 10 \times 10 \times 10 \times 10 \times 10 \times 10 \times 10$

$= 10 \times 10 \times 10 \times 10 \times 10 \times 10 \times 10 \times 10 \times 10$

$= 1\,000\,000\,000$ OR 10^9

This method is tedious but it has been shown that:

$10^2 \times 10^3 = 10^5$ and $10^4 \times 10^5 = 10^9$

Examine the index figures 2 + 3 = 5, 4 + 5 = 9. To multiply the index figures are added so:

$10^2 \times 10^3 = 10^{2+3} = 10^5$

This method saves the trouble of writing all the noughts.

Remember It is only possible to carry out multiplication by adding the index figures where the base is the same:

$10^5 \times 10^6 = 10^{11}$

$7^2 \times 6^4 = \cancel{13^6}$ Base not the same: this method cannot be used.

Examples (answers at the end of the chapter)

60. $10^6 \times 10^4$

61. $6^{10} \times 10^6$

62. $5^2 \times 5^3$

63. $10^1 \times 10^1$

Division

Division can also be dealt with by using a similar method, for example:

$$10^5 \div 10^2 = \frac{10^5}{10^2}$$

$$\text{OR} \quad \frac{10 \times 10 \times 10 \times 10 \times 10}{10 \times 10}$$

$$= \frac{100\,000}{100}$$

$$= \underline{\underline{1000}}$$

Rewrite 1000 as a base and index $= 10^3$

Is there any connection between $10^5 \div 10^2 = 10^3$

The index numbers have been *subtracted*

$$10^9 \div 10^6 = \frac{10^9}{10^6} = 10^3$$

Examples

1. $10^6 \div 10^4 = 10^2$
2. $10^3 \div 10^2 = 10^1$

It has been discovered that $10^5 \div 10^2 = 10^{5-2} = 10^3$

Suppose the question is rewritten as $10^5 \times 10^{-2} = 10^3$

The answer is the same: $10^5 \div 10^2$ OR $\frac{10^5}{10^2}$ OR $10^5 \times 10^{-2}$

All three ways of expressing this problem are correct. Examine the following problem and remember the rules for division when dealing with fractions, invert and multiply.

$$10^5 \div 10^3 = \frac{100\,000}{1000}$$

$$\text{OR} \quad 100\,000 \div 1000 = \frac{100\,000}{1} \div \frac{1000}{1} \quad \text{OR} \quad \frac{100\,000}{1} \times \frac{1}{1000}$$

Note that $\frac{1}{1000} = \frac{1}{10^3}$. To carry out division the index is subtracted so 10^3 becomes effectively 10^{-3}, so that

$$\frac{100\,000}{1} \times \frac{1}{1000} = \frac{10^5}{10^3} \quad \text{OR} \quad 10^5 \times 10^{-3} = 10^2$$

In this example dividing by a positive index value is exactly the same as multiplying by a negative index, the rule to remember is very simple. If the position of a base and index is changed, i.e. from dividing (below line), to multiplying (above line), change the sign of the index.

Examples

$\frac{10^6}{10^3} = 10^6 \times 10^{-3}$ ← Change from below line to above. 10^3 becomes 10^{-3}

$\frac{10^6}{10^{-3}} = 10^6 \times 10^3$ ← Change from below line to above. 10^{-3} becomes 10^3

Consider the next problem $10^2 \div 10^6$

To divide the index number is subtracted $10^{2-6} = 10^{-4}$

Check using long hand:

$$\frac{10^2}{10^6} = \frac{100}{1\,000\,000} = \frac{1}{10\,000} = \frac{1}{10^4} = 10^{-4}$$

What is the answer to the following:

$10^{-4} \div 10^2$

Standard form

The manipulation of indices can be extremely useful when used in conjunction with another form of shorthand called *standard form.* This is a neat way to write large or small numbers and complements the index exercises already completed. It has already been proved that $100 = 10^3$. Other examples would be:

$2000 = 2 \times 1000 = 2 \times 10^3$

$200 = 2 \times 100 = 2 \times 10^2$

$20 = 2 \times 10 = 2 \times 10^1$

$2 = 2 = 2 \times 10^0$

Any number can therefore be put into standard form. All that is happening is that the decimal point is being manipulated and the power of the index is used as an indicator of the amount of movement:

$2000 = 2000. = 2. \times 10^3$

The decimal point has moved *three places* to the *left* so the index is 3 and *positive*. Can a number less than 1 be written in standard form?

$0.0582 = 5.82 \times 10^{-2}$

The decimal point has been moved *two places* to the *right* so the index is *negative*. The sign of the index tells us which way the decimal point has been moved. In standard form the value is shown as 5.82×10^{-2}, the -2 says that the decimal point has been moved two places to the right. To return the number to normal form the decimal point has to be moved back to its original position, it must be moved two places to the left:

5.82×10^{-2}	Standard form
$= 0.0582$	Original form

Put the following into standard form:

1. $269.54 = 2.6954 \times 10^2$
2. $3011.4 = 3.011 \times 10^3$
3. $6.56 = 6.56$
4. $0.0721 = 7.21 \times 10^{-2}$
5. $0.123 = 1.23 \times 10^{-1}$

Change the following from standard form to true values (answers at the end of the chapter):

64. $6.23 \times 10^2 =$
65. $7.791 \times 10^1 =$
66. $9.954 \times 10^6 =$
67. $5.25 \times 10^{-2} =$
68. $6.314 \times 10^{-3} =$

Using your calculator

Calculation will be achieved quickly and accurately if you use a calculator; however, different manufacturers have different concepts of which functions are most important so keys are placed in different positions. Some keys may have two or more functions. You *must* keep your instruction booklet that came with your calculator. The keys that are used most often are shown below:

The normal maths functions	$+ - \times \div$
The decimal digits	0–9
The decimal point	.
The equals sign	=
EXP	
+/–	
Square root	√
Reciprocal function	1/x

Most people can already use the basic functions, $+ - \times \div$, so their use will be taken as understood. It will probably have been discovered by this stage that in electronics it is necessary to deal with a large range of values, e.g. 10 MΩ, 150 kΩ, and very small values, e.g. 100 millihenries, 25 microamps. In calculations these values are normally written as a numerical value raised to a power of ten (i.e. scientific notation). A 10 kΩ resistance used in a calculation would be reduced to a value of 10 but with power of 10^3, 10 kΩ $= 10 \times 10^3$.

There are some standard terms which are commonly used in engineering:

mega	1 000 000	M	10^6
kilo	1000	k	10^3
milli	$\frac{1}{1000}$	m	10^{-3}
micro	$\frac{1}{1\,000\,000}$	μ	10^{-6}
nano	$\frac{1}{1\,000\,000\,000}$	n	10^{-9}
pico	$\frac{1}{1\,000\,000\,000\,000}$	p	10^{-12}

To enter the value 10 kΩ as a mathematical quantity for use in a calculation the keys on the calculator are pressed in the following order, check your calculator instruction booklet:

10 EXP 3

To enter the value 10 mA it is necessary to remember that 10 mA has a scientific notation of 10×10^{-3}. Now press the following keys:

10 EXP +/– 3

Notice when you press the EXP key the display changes; again it is difficult to say exactly what is seen because there is no uniformity amongst calculator manufacturers, but typically the display may show:

10^{00} or 10 ↑ 00
space

When entering 10 mA (10 EXP +/− 3), the display will go to:

10^{-03} or 10 ↑ −03
space

Consider the following examples and use your calculator to determine the results:

1. A voltage of 300 volts will apear across a 100 kΩ resistor when . . . of current is flowing. Determine the value of the current. The formula to use is Ohm's law.

$$I = \frac{V}{R}$$

$$I = \frac{300}{100\ \text{k}\Omega}$$

Carry out this calculation on your calculator by pressing the following keys:

300 ÷ 100 EXP 3 =

The display should now show $3.^{-03}$.

Now convert the calculator display into a current remembering that the −03 signifies milli so the answer is 3 mA.

2. Determine the time constant of the following circuit:

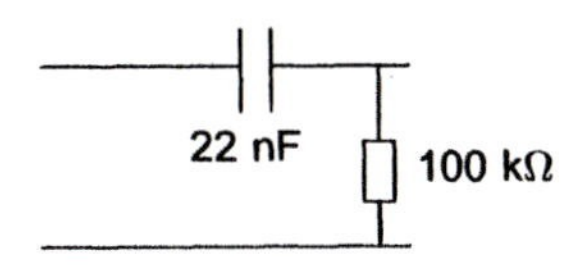

Time constant $t_c = C \times R$

$t_c = 22\ \text{nF} \times 100\ \text{k}\Omega$

Press the calculator keys in the following order:

22 EXP +/ − 9 × 100 EXP 3 =

The answer displayed should be 2.2^{-03}.

Since time is measured in seconds the answer is 2.2 milliseconds, i.e. 2.2 ms.

Repeat this calculation, but change the value of resistor so that it is now only 1 kΩ.

$t_c = C \times R$

$= 22 \text{ nF} \times 10 \text{ k}\Omega$

Press the following keys:

22 EXP +/ − 9 × 10 EXP 3 =

The result is shown as 2.2^{-04}. The −04 does not fit with the standard terms of engineering so how can it be changed? If your calculator has an ENG key, press it. The display should now change to $220.^{-06}$. Using scientific notation the −06 represents the term micro, so the answer is now 220 μs.

Consider a circuit where three resistors are connected in *parallel*, $R_1 = 5.6$ kΩ, $R_2 = 6.8$ kΩ and $R_3 = 3.3$ kΩ. The normal method of calculation would be by using the formula below and applying the rules for adding fractions:

$$\frac{1}{R_T} = \frac{1}{R_1} + \frac{1}{R_2} + \frac{1}{R_3}$$

$$= \frac{1}{5.6\text{k}} + \frac{1}{6.8\text{k}} + \frac{1}{3.3\text{k}}$$

The calculator can be used to perform the addition of these fractions. All values are in kΩ so ignore the k at this point and the calculation becomes:

$$= \frac{1}{5.6} + \frac{1}{6.8} + \frac{1}{3.3}$$

To perform this calculation press the following keys:

$$5.6\frac{1}{x} + 6.8\frac{1}{x} + 3.3\frac{1}{x} = \frac{1}{x}$$

$= 1.59$

To use the $\frac{1}{x}$ function it may be necessary to press another key on your calculator first, check your instruction booklet.

Remember the k ignored earlier, reinsert it, so that the answer now becomes 1.59 kΩ.

Numbering systems

The decimal system

This is the system used in everyday transactions and the system used to *display* answers on a calculator. However, the decimal system is not the system used by the calculator to perform its functions. Like many of today's electronic devices the calculator uses digital circuitry which relies on a binary signalling system. An understanding of numbering systems is important. There are many operations carried out in calculations without the realisation that rules are being applied. Examining the facts of the decimal system reveals the following.

The system has a *radix* or *base* of 10. This means it uses ten different digits to represent values 0, 1, 2, . . ., 9. It also uses *positional notation,* which is the mathematicians way of saying that the position of a digit determines its value.

e.g. 649. ← decimal point

The figure on the left, 6, is a digit between 0 and 9 but in this case its value is not simply 6, but 600. The hundreds is denoted by its position relative to the *radix point* commonly referred to as the decimal point. The value/position of the number can be shown as below where a digit's value is shown raised to a power of the base number of the system used:

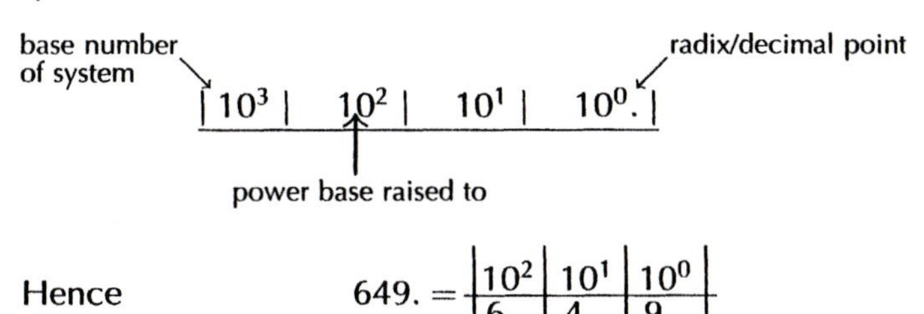

$10^2 = 10 \times 10 = 100$

Hence 6 lots of $10^2 = 6 \times 100 = 600$

$10^1 = 10 \times 1 = 10$

Hence 4 lots of $10^1 = 4 \times 10 = 40$

$10^0 = 10^0 = 1$

Hence 9 lots of $10^0 = 9 \times 1 = 9$

$649 = 600 + 40 + 9$ expressed in simplest terms

All numbering systems make use of these rules. The only change being the base – the number of digits used.

The binary system

It has been stated earlier that a calculator displays the answer using the decimal system; however, it did not use this system to perform the calculations, but used a digital/binary signalling system. An example of binary signalling is where 0 volts represents one digit and a higher level of voltage, e.g. 5 volts, is the other. This system uses two digits and is therefore called the binary system. The digits used are one and zero. It is convenient that all the other facts are the same as the decimal numbering system. The base of the binary system is 2 and only two digits are used. Positional notation is used exactly as before. A binary number shown as 1011. has a value, in decimal terms, of 11 (eleven). Using a table of positional notation the value can be calculated:

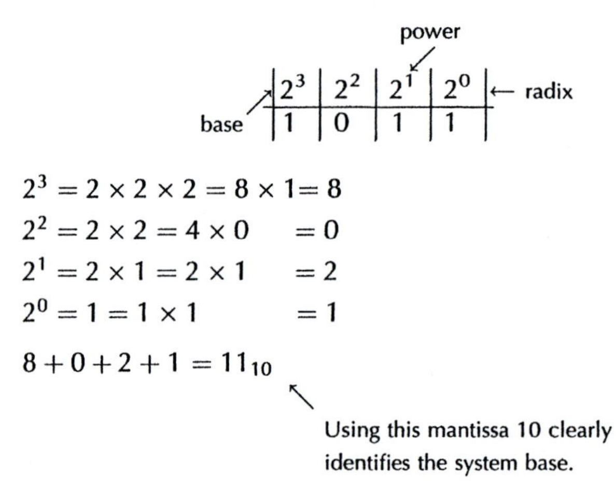

Using a table again, consider another binary number and determine its decimal value,

e.g. 100101_2 (identifies number as binary):

2^5	2^4	2^3	2^2	2^1	2^0.
1	0	0	1	0	1

It is worthwhile at this stage to remember the table in another form showing the decimal value of each binary digit according to its position:

32	16	8	4	2	1
2^5	2^4	2^3	2^2	2^1	2^0.
1	0	0	1	1	1

$= 32 + 4 + 1 = 37_{10}$

The number on the left-hand side is called the most significant bit (MSB) because it is the largest value, while the number on the right-hand side is known as the least significant bit (LSB) because it has the smallest value. To change a decimal number from its base of ten to a new base the decimal number is divided by the new base required and the remainder noted at each step. For example, to convert from decimal (base ten) to binary (base two) successive division by two is used. Converting 29_{10} to its binary equivalent is achieved as follows:

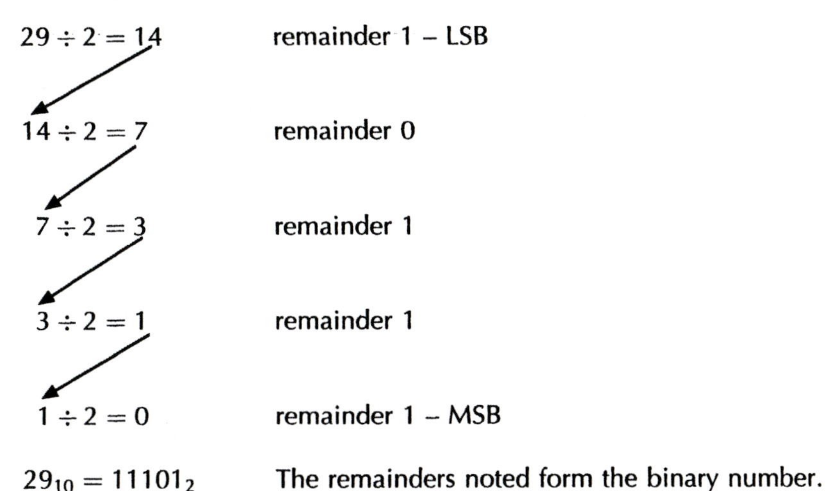

Consider another example converting 35_{10} to its binary equivalent:

$35 \div 2 = 17$	remainder 1 – LSB
$17 \div 2 = 8$	remainder 1
$8 \div 2 = 4$	remainder 0
$4 \div 2 = 2$	remainder 0
$2 \div 2 = 1$	remainder 0
$1 \div 2 = 0$	remainder 1 – MSB

MSB ↘ ↙ LSB

$35_{10} = 100011_2$

Examples

Convert the following into their binary equivalents (answers at the end of the chapter):

69. 31_{10} 70. 51_{10} 71. 84_{10} 72. 130_{10}

Having converted these numbers is there any interesting poinst to note in the answers? Convert the following binary numbers into their decimal equivalents:

73. 10011_2 74. 111011_2 75. 01110_2 76. 100010_2

Again having converted these numbers are there any interesting points to note in the answers? Odd numbers always have a binary 1 as their LSB and even numbers a binary 0 as their LSB.

Other numbering systems

Two other numbering systems that are encountered in electronics are the *octal* system with a base of 8 and the *hexadecimal* system with a base of 16. Both these systems are convenient methods to use because of their ease of conversion.

The octal system

This system uses eight digits 0 to 7 – hence its base of 8. It makes use of positional notation in exactly the same way as the decimal or binary systems. It is also a helpful way of remembering binary codes. For example, consider the binary number 110110001. With such a chain of digits it is easy to miss out a digit which will of course change its value. In its octal equivalent this binary number could be stated as 661_8 which is much simpler to remember or write. Three digits instead of nine. Converting the binary to decimal could be employed, but this is not quite as easy. To convert from binary to octal the binary word is divided into groups of three starting at the right-hand side.

Hence

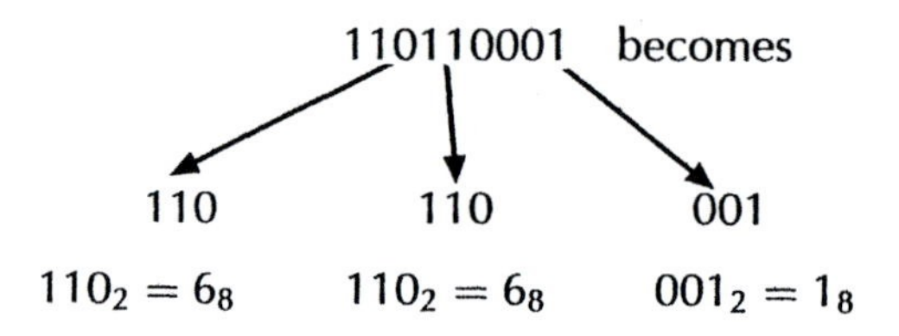

$110_2 = 6_8$ $110_2 = 6_8$ $001_2 = 1_8$

To be able to do this it is necessary to remember each of the binary codes with three bits for each of the octal numbers:

$0_8 = 000_2$ $48 = 100_2$

$1_8 = 001_2$ $5_8 = 101_2$

$2_8 = 010_2$ $6_8 = 110_2$

$3_8 = 011_2$ $7_8 = 111_2$

Although octal numbers look very much like decimal numbers their value is very different. Again employing a table which uses positional notation, the decimal equivalent of 661_8 can be determined:

8^2	8^1	8^0.
6	6	1

$8^2 = 8 \times 8 = 64 \quad 64 \times 6 = 384$

$8^1 = 8 \times 1 = \ \ 8 \quad \ 8 \times 6 = \ \ 48$

$8^0 = 1 \times 1 = \ \ 1 \quad \ 1 \times 1 = \ \ \ 1$

It is worth remembering that any number raised to the power zero = 1:

$384 + 48 + 1 = 433_{10}$

Compare this answer with one obtained when the binary number is converted directly into decimal using a conversion table:

512	256	128	64	32	16	8	4	2	1
2^9	2^8	2^7	2^6	2^5	2^4	2^3	2^2	2^1	2^0
	1	1	0	1	1	0	0	0	1

$= 256 + 128 + 32 + 16 + 1 = 433_{10}$

Hence $110110001_2 = 661_8 = 433_{10}$

These three methods are all acceptable for representing numerical quantities.

Hexadecimal system

This is another convenient system for representing binary numbers or chains of binary information. The hexadecimal system is based on sixteens. It may be referred to as HEX and given the symbol H. It is different to the numbering systems used so far because it requires the use of alphanumerics for certain values. It is necessary to remember the sixteen codes and their binary equivalents as shown below:

Decimal	HEX	Binary
0	0	0000
1	1	0001
2	2	0010
3	3	0011
4	4	0100
5	5	0101

Decimal	HEX	Binary
6	6	0110
7	7	0111
8	8	1000
9	9	1001
10	A	1010
11	B	1011
12	C	1100
13	D	1101
14	E	1110
15	F	1111

The use of the letters A to F is required to distinguish between ten – 10_{10} – and one followed by a zero – $1_{10}0_{10}$:

Ten = $10_{10} = 1010_2 = A_{16}$ or AH

One Zero = $10 = 0001_2000_2 = 10_{16}$ or 10H

Its use is in microprocessors for representing binary codes. For example, the HEX code $3E_{16}$ tells a Z80 microprocessor to load its accumulator. While it is easy to remember 3E, it is not so easy to remember 00111110_2. It is possible to apply all the same principles previously used to this numbering system. To convert 48_{10} to hexadecimal it is divided by the base of the system required and the remainder noted at each step:

$48 \div 16 = 3$ remainder 0 – LSB

$3 \div 16 = 0$ remainder 3

Therefore $48_{10} = 30_{16}$ or 30H.

To convert 30H to decimal draw a table:

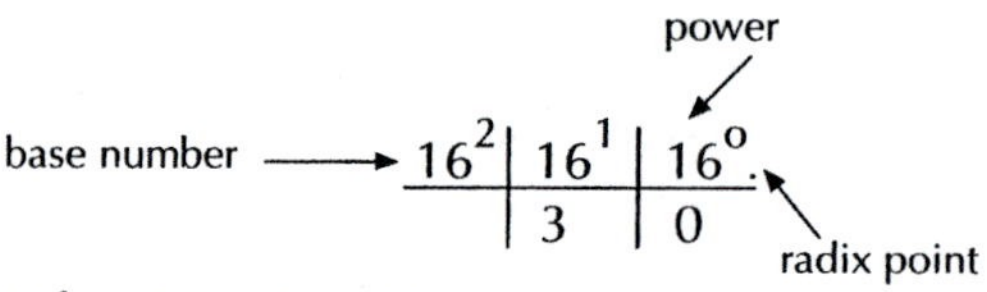

power

base number → 16^2 | 16^1 | 16^0.

16^2	16^1	16^0.
	3	0

radix point

$16^1 = 16 \times 1 = 16 \times 3 = 48$

$16^o = 1 = 0 \times 1 = 0$

Hence 30H = 48_{10}

Conversion between hexadecimal, octal and binary is also very convenient provided that the codes shown above have been remembered:

30H = 00110000_2

To convert the 8 bit binary code above into octal the binary digits must be rearranged into 3 bit groups starting to group on the right. There is a problem in so much as to rearrange the bits into groups of three is not possible without adding an extra bit on the left so that the binary code becomes .000 110 000. It now has nine bits but retains its original value. To convert 000 110 000 to octal using the remembered 3 bit codes results in 060_8.

Hence $000110000_2 = 060_8$

Adding binary numbers

Numbers in binary form may be added in the same way as decimal numbers by obeying the following rules:

	Sum	Carry
0 + 0 =	0	0
0 + 1 =	1	0
1 + 0 =	1	0
1 + 1 =	0	1

To add the following binary numbers 1011_2 and 1000_2, place the numbers as below so that the LSBs on the right are always in line:

Column	5	4	3	2	1
		1	0	1	1
		1	0	0	0

Now add column 1 applying the rules shown.

1. 1 + 0 = 1 sum and 0 carry

Now add column 2 applying the rules shown.

2. 1 + 0 = 1 sum and 0 carry

Now add column 3 applying the rules shown.

3. 0 + 0 = 0 sum and 0 carry

Now add column 4 applying the rules shown.

4. 1 + 1 = 0 sum and 1 carry

The answer appears to be 0011 but there is a carry generated which must be carried forward into column 5, the next available column. Since there are no digits shown in this column it is understood that they are both zero, so the carry is added to zero and becomes 1.

5. 0 + 0 + 1 carry = 1 sum and 0 carry

The result of adding 1011_2 to 1000_2 is 10011_2. The answer for simple numbers can always be checked by converting to decimal and comparing the answers. In the examples shown 11_{10} is added to 8_{10} which of course gives a result of 19_{10}. Converting this figure to binary produces the result 10011_2 as before.

The system works but is a little clumsy. However, it is usual to find a binary addition where it is only necessary to add two digits because of the generation of the carry, so it is helpful to remember the following rules:

RULE	CARRY +	DIGIT 1 +	DIGIT 2 =	SUM	CARRY
1.	0	0	0	0	0
2.	1	0	0	1	0
3.	0	0	1	1	0
4.	1	0	1	0	1
5.	0	1	0	1	0
6.	1	1	0	0	1
7.	0	1	1	0	1
8.	1	1	1	1	1

By remembering the above rules binary addition is simpler.

Adding 10111_2 to 10011_2

10111
10011

Starting with least significant bits on the right-hand side. Apply rule 7:

1. Carry = 0 Digit 1 = 1 Digit 2 = 1 = Sum = 0 Carry = 1
   Write 0 in the LSB position in your answer.

2. Add the bits in the next column and apply rule 8. 10111
   10011

   Carry = 1 Digit 1 = 1 Digit 2 = 1 = Sum = 1 Carry = 1

   Write 1 in the next LSB position in your answer. Repeat the process until all bits are added and the resultant will be 101010_2.

Answers to examples

1. 15 2. 12 3. 21 4. 54 5. 40

6. $\frac{11}{24}$ 7. $\frac{121}{168}$ 8. $\frac{31}{60}$ 9. $\frac{43}{55}$ 10. $\frac{37}{32}$

11. $1\frac{1}{3}$ 12. $3\frac{1}{3}$ 13. $8\frac{1}{2}$

Additions

14. $3\frac{5}{16}$ 15. $11\frac{17}{40}$ 16. $14\frac{1}{8}$ 17. $10\frac{5}{12}$ 18. $8\frac{13}{14}$

Subtractions

19. $\frac{5}{12}$ 20. $\frac{1}{2}$ 21. $\frac{2}{3}$ 22. $\frac{55}{69}$ 23. $\frac{19}{24}$

24. $\frac{9}{16}$ 25. $2\frac{11}{16}$ 26. $2\frac{1}{12}$ 27. $3\frac{23}{24}$ 28. $1\frac{5}{24}$

29. $1\frac{1}{2}$ 30. $\frac{-13}{16}$ 31. $1\frac{7}{16}$ 32. $1\frac{3}{4}$ 33. $1\frac{13}{16}$

34. $\frac{1}{3}$ 35. $\frac{1}{2}$ 36. $\frac{12}{77}$ 37. $\frac{6}{35}$ 38. $\frac{7}{18}$

39. $\frac{2}{3}$ 40. $\frac{2}{3}$ 41. $\frac{5}{8}$ 42. $\frac{24}{35}$ 43. $\frac{4}{5}$

44. $9\frac{7}{12}$ 45. $6\frac{2}{3}$ 46. 3 47. $\frac{5}{8}$ 48. $2\frac{5}{8}$

49. 34 50. $2\frac{1}{2}$ 51. 830 52. 0.026 53. 79

54. 7.7 55. 1 56. 29 57. 10.8 58. 8.1

59. 3.3 60. 10^{10} 61. Base not the same 62. 5^5

63. 10^2 64. 623 65. 77.91 66. 9954000

67. 0.0525 68. 0.006314 69. 11111_2

70. 110011_2 71. 1010100_2

72. 10000010_2 73. 35_{10} 74. 59_{10}

75. 14_{10} 76. 34_{10}

4

Science background

In this chapter four science elements, *Heat, Mechanical Units, Light,* and *Sound,* will be examined. The aim is not to embark upon a study of physics to 'A level' standard, but rather to examine some fundamental areas of science that relate directly to electronics servicing. Like the previous chapter, this chapter should be used as a source of reference.

Heat

The effects of heat are used extensively in electrical and electronic circuits and devices. For example, heating elements are used in room heaters, kettles, cookers, etc. Similarly, a heater element is essential to stimulate thermionic emission in thermionic valves; two common examples being the cathode ray tube and the vacuum fluorescent display used in VCRs, hi-fi, microwave ovens, etc. The magnetron used to generate the microwave energy in an oven is also a thermionic valve. As we shall see in Chapter 6, the thermistor is a component that makes use changes in temperature.

Heat can also be an enemy of electronic circuits. Early radio and television receivers employing large numbers of valves were very unreliable. This was partly due to the fact that they operated at a very high temperature (generated by the valves) which resulted in component failure. There are still components today that have a very high operating temperature, and if this is not compensated for, rapid failure of components will result.

There are three ways in which heat moves: *conduction, convection,* and *radiation.*

Conduction is the movement of heat through a material. In practice the material is a solid or a liquid; gases are not generally good conductors of heat. Some solids are better conductors of heat than others, the most efficient conductors being those that have large numbers of free electrons, which also makes them electrical conductors. However, there are some electrical insulators that conduct heat efficiently, and some of these materials find their uses in circuit construction.

Convection occurs when a liquid or gas (both classed as fluids) is heated. The heating effect causes the fluid to expand which means that its density decreases. When influenced by a gravitational field the less dense (heated) fluid rises and the cooler fluid sinks. If the source of heat is at the bottom of the container in which the fluid is held, then as the cooler fluid sinks then it becomes heated, causing it to rise. Thus a circulating movement of the fluid occurs within the container.

Radiation of heat is fundamentally different from conduction and convection because it is not essential for the heat to move through any medium. In other words, radiation can take place through a complete vacuum. Radiated heat is *electromagnetic radiation* in the infra-red region which lies just outside the visible light spectrum (see Figure 4.4). Being by nature electromagnetic radiation, radiated heat behaves very much the same as light or radio waves, which will be examined later.

Any matter with a temperature that is above absolute zero (the point at which there is no heat at all, −273°C) will emit infra-red radiation. For example, even a block of ice with a temperature of 0°C emits infra-red heat; however, it will still melt if brought into the path of heat radiated from a fire because the amount of radiation absorbed by the ice will now be greater than the amount given off, and its temperature will thus rise.

Just as visible light reflects off a surface, so do heat rays. And like visible light, brighter surfaces reflect to a greater degree than dull surfaces. A dull surface absorbs the heat. On the other hand, surfaces that are good reflectors of heat are not good emitters of heat; those that are good absorbers are also good emitters.

One common electronic device that makes use of radiated heat is the passive infra-red intruder detector. The heat that is radiating from every item in the room is focused onto a quartz crystal, heating it to a certain temperature. When an intruder enters the room, the heat from his body causes an increase in the crystal temperature, the crystal thus expands and in doing so produces an electrical voltage which can be used to trigger the detector. Even if the temperature of the intruder is lower than that of the surrounding area, the device will still trigger because the crystal will cool causing it to contract.

It was stated earlier that heat can be an enemy of electronic components or circuits. The most common example of this is associated with semiconductors.

At absolute zero a semiconductor is an insulator, but as its temperature is increased it conducts electricity. For example, consider the case of a transistor. The higher its temperature the lower its resistance becomes, but this causes more current to flow which in turn increases its temperature. The transistor is now caught in a cycle where its temperature continues to rise, causing its resistance to fall, causing increases in current, which in turn raises its temperature. A point is reached where the silicon crystal is so hot that it breaks down, and the transistor is destroyed. This effect is known as *thermal runaway.*

All semiconductor devices are prone to thermal runaway, and for this reason many transistors, ICs, and diodes are mounted onto aluminium heat sinks. The principle is illustrated in Figure 4.1 where a power transistor with its own integral metal heat sink is shown bolted to a flat aluminium plate. *Heat is removed from the transistor by conduction,* the heat travelling through the heat sink. It is dissipated from the heat sink by both *convection* through the air and *radiation.* To increase the radiated output, the heat sink is often coloured black; remembering that a dull surface is a better emitter of heat.

Figure 4.1

Power transistor with heat sink

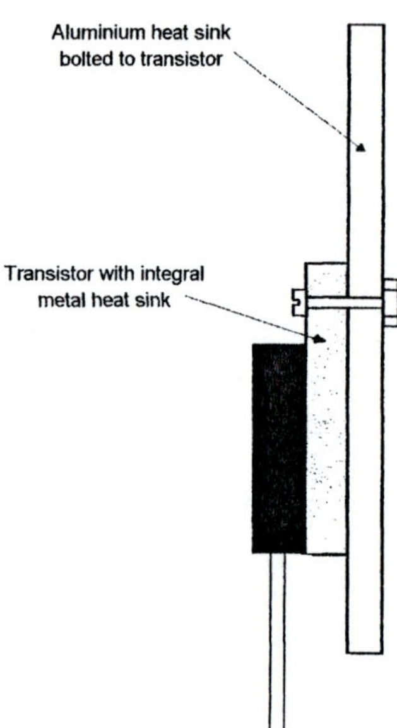

In equipment where the temperature of the heat sinks is exceptionally high, such as large power amplifiers and computers, a fan may be employed to increase the air flow across the heat sink. Another way of improving the air flow is to have fins on the heat sink (Figure 4.2).

Figure 4.2
Examples of heat sinks

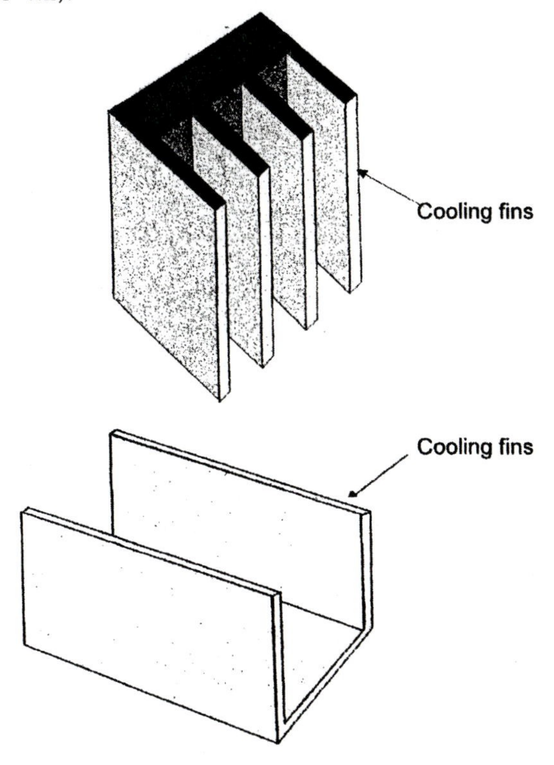

The efficiency of the conduction between the device and its heat sink is impaired by the fact that the two metal surfaces may not make good physical contact because of minute irregularities – even though they may appear smooth to the naked eye. For this reason a silicon compound is often applied to the device before it is fixed to the heat sink. This compound has good heat conductive properties, and a very fine composition so that it fills in the irregularities, thus improving the heat transfer between the two surfaces. When replacing components that use heat sink compound, it is essential that the old compound is removed if it has dried up, and new compound applied, otherwise premature failure of the new component may occur.

When replacing devices that are mounted onto heat sinks, take care to refit any mica insulating plates that may be placed between the device and the heat sink. These are included to provide electrical insulation between the device and the heat sink, which may be essential if more than one device is fitted to the same heat sink.

Mechanical units

In physics, *mass* is defined as (a) the amount of matter that a body contains and (b) the measure of a body's resistance to change in motion (its inertia).

All too frequently mass and weight are considered to mean the same thing, but in reality this is not so. The weight of a body is dependent upon the force of gravity acting upon it, so a body with a certain mass will weigh differently under differing gravitational forces. This is clearly illustrated when a man is observed walking on the moon where he is quite clearly lighter than he is on the earth, yet his mass remains unchanged.

Mass is measured in kilograms (symbol kg). Perhaps this is why there is some confusion between mass and weight. The unit of the kilogram was defined by taking a precise mass which happens to be a cylinder of platinum–iridium. This alloy cylinder is maintained at a constant temperature and is held at Sèvres in France. When in the influence of the earth's gravitational field this mass is taken to be 1 kg.

Weight is a measure of the *force* acting upon a mass. Force is defined as being any action or influence that accelerates, or changes the velocity of, an object. *The unit of force is the newton* (symbol N), named after Sir Isaac Newton, the seventeenth-century physicist and mathematician. This unit is defined as that force which imparts an acceleration of 1 m/sec^2 on a mass of 1 kg.

So, if a mass of 1 kg is taken and the force acting upon it at the surface of the earth in Sèvres is measured, its weight is 9.81 N/kg (newtons per kilogram). Of course, when measured elsewhere such as on the moon its weight will be different. Even when measured at different points on the earth the weight will vary, because the gravitational force is not constant.

From this we can derive the formula:

$$\text{Weight} = \text{Mass} \times \text{Force (of gravity)}$$

In physics, *work* is defined as the product of a force applied to a body, and the displacement of the body in the direction of the applied force.

In order to move a mass through any distance a certain amount of energy has to be expended. It can be considered that this energy is transferred into the mass. For example, if a box is lifted onto a table then the energy used to lift the box can be considered to be transferred to the box, because it now contains an equal amount of energy to move itself back down. This would be evident if you pull the table from under it!

The unit of work is the joule (symbol J), which is the work done by a force moving the mass on which it is acting through a distance. When a force of 1 newton moves the object on which it is acting through a distance of 1 metre, then the work done is said to be 1 joule (1 J). The latent energy held in the mass of the box in our example above is also measured in joules.

From the above arguments we can derive the formula:

Work done = Force × Distance

where force is measured in newtons and distance is measured in metres.

Summarising

Mass is measured in kilograms (kg)
Force is measured in newtons (N)
Weight (being a function of force) is measured in newtons (N)
Work done is measured in joules (J)

Light

Light and radio waves are both forms of electromagnetic radiation, the main difference being their frequencies. Having studied the ways in which light travels through free space (propagates), and the effects of reflection and refraction, it will be discovered that very similar principles apply to the propagation of radio waves.

Although the world's leading physicists are still uncertain about the true nature of light, they are generally of the opinion that it is made up from minute particles called photons. Light also appears to travel in sinusoidal waves, and the frequency of these waves determines the colour. The velocity of both light and radio waves in a vacuum is approximately 300 000 000 metres per second, i.e. 300×10^6. This means that each different frequency must have a corresponding wavelength, which is illustrated in Figure 4.3 where the lower frequency wave A will travel much further from the source during the period of one cycle than the higher frequency wave B. The distance travelled by one cycle of a wave is known as the wavelength, which is measured in metres and denoted by the Greek symbol λ (lambda).

From Figure 4.3 we can see that the wavelength of signals becomes shorter as the frequency increases. The relationship between frequency and wavelength can be expressed as

$$f = \frac{v}{\lambda}$$

where f = frequency, v = velocity of propagation (300×10^6 m/sec) and λ = wavelength.

Figure 4.3

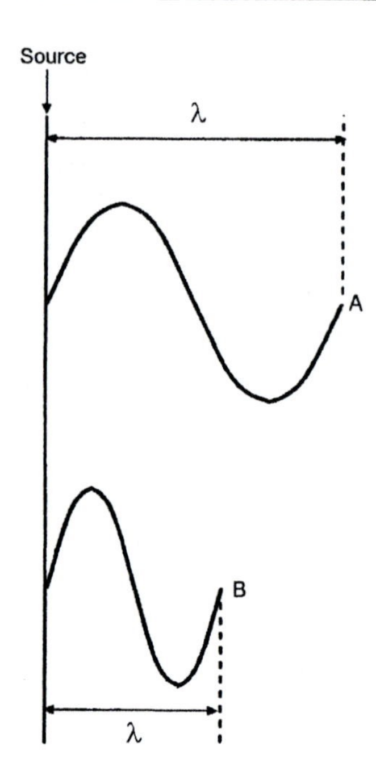

For example, taking a wavelength of 650 nm, the frequency of red light can be found:

$$f = \frac{3 \times 10^8}{650 \times 10^{-9}} = 0.46 \times 10^{15} = 460 \text{ Tera Hz}$$

Moving towards infra-red the frequency reduces, and towards ultra-violet the frequency increases.

The known frequency range of the electromagnetic spectrum is shown in Figure 4.4 where we can see the range of frequencies used for radio transmission, and the much higher, but somewhat narrow, range of frequencies to which the human eye responds. Also illustrated in this figure are the range of colours which make up the visible light spectrum, i.e. Red, Orange, Yellow, Green, Blue, Indigo, Violet, frequently memorised using the phrase Richard Of York Gave Battle In Vain.

The eye reacts to light, converting the incoming electromagnetic radiation into small electrical signals which are sent to the brain. The brain converts these signals into an image.

The eye contains four sets of cells; one set has a cylindrical structure and is known as a rod. The other three sets are conical in shape and are known as cones. The cones are sensitive to different frequencies of light, and it is these which enable the eye to differentiate between colours.

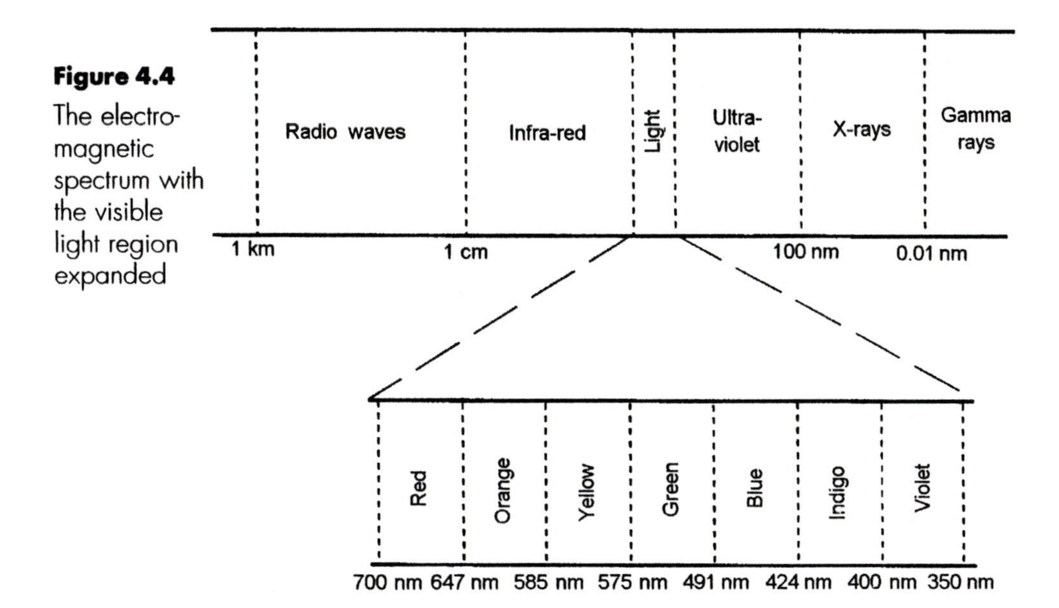

Figure 4.4
The electro-magnetic spectrum with the visible light region expanded

One set of cones responds to the range of frequencies with corresponding wavelengths in the order of 600–700 nm (nanometres). Upon receiving signals from these cones the brain acknowledges that it has seen red light. The second set of cones responds to wavelengths around 500–600 nm, that is, green light, and the third set responds to 400–500 nm, blue light.

The eye does not respond equally to all frequencies, nor is the response equal for all people. As Figure 4.5 illustrates, the response is best at wavelengths around 550 nm; that is, green light.

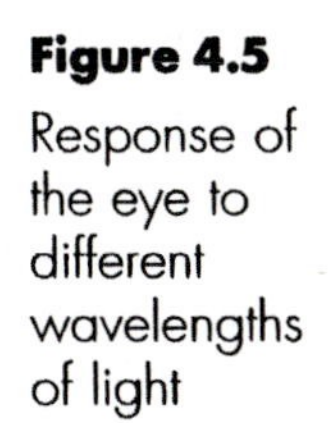

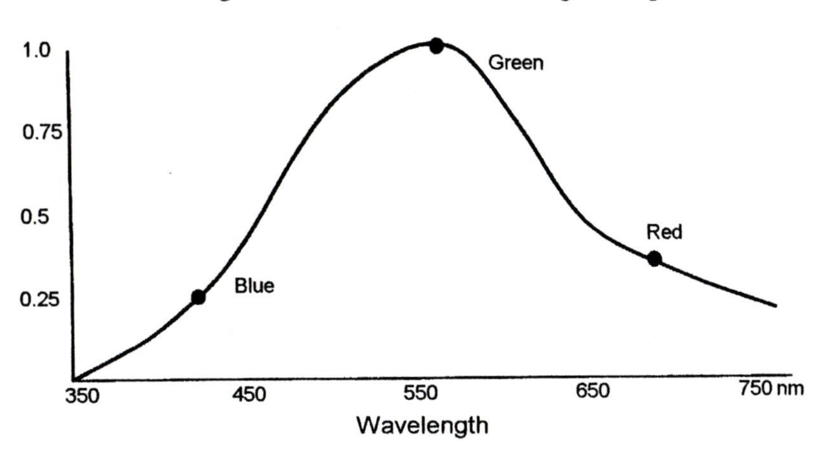

Figure 4.5
Response of the eye to different wavelengths of light

What we perceive to be white light is actually red, green and blue light being emanated simultaneously from a source. These colours are known as *primary colours*. The brain is constantly integrating the information from the cones in the eye, producing three resultant *secondary colours* from the three primaries. This process of *additive mixing* is illustrated in Figure 4.6.

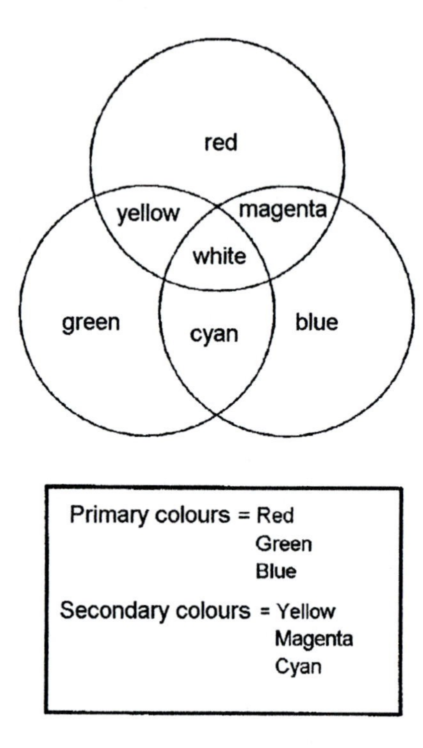

Figure 4.6
Principle of additive mixing

The rods are not frequency (and therefore colour) conscious. These cells measure the intensity of the total amount of incoming radiation. Thus they determine if the scene is very bright, very dark, or somewhere in between. If the eye were to contain only rods we would see everything in black and white.

Summarising, the rods determine the black and white content of an image, and the cones determine the colour content. The brain constantly processes the information coming from the four sets of cells to determine the brightness and colour of the image. This basic understanding of how the eye functions is of particular importance when we come to consider the operation of a colour television, which to all intents and purposes is a device designed to fool the eye into believing that it is observing moving coloured images.

As far as we are concerned, we can assume that light travels in a straight line through a vacuum. However, once it strikes a surface it will reflect, although not all frequencies will reflect off a given surface, some may be absorbed.

The way in which light reflects off a surface is very predictable, and can be explained using the experiment illustrated in Figure 4.7. From this experiment two important points relating to light reflection can be demonstrated. First, that the incident ray, the reflected ray, and the normal (a straight line drawn out from the point of incidence at the mirror) all lie in the same plane. Second, the angles of the incident ray and the reflected ray are always the same.

Figure 4.7
θ_i = angle of incidence and
θ_r = angle of reflection

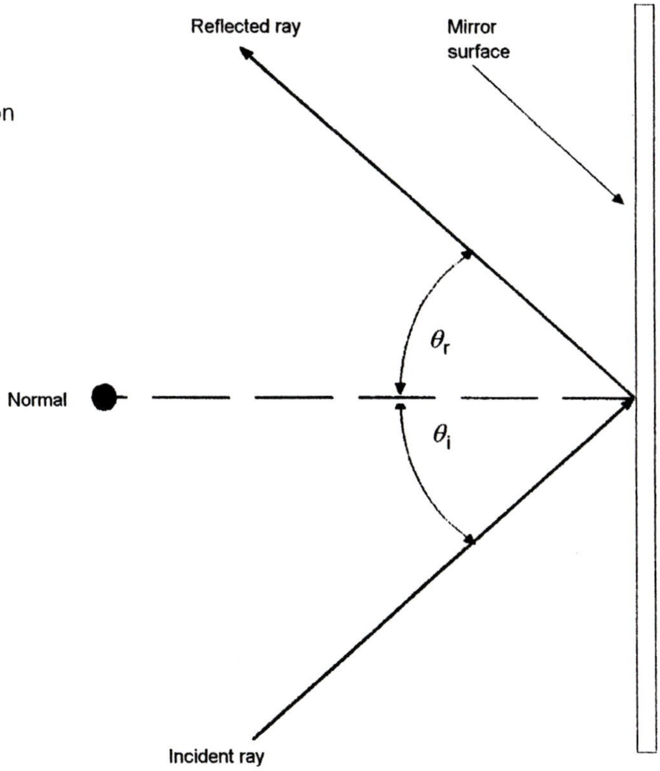

Reflected ray
Mirror surface
θ_r
Normal
θ_i
Incident ray

Refraction is the effect solid materials have on light travelling through them, and is more complex than reflection. Light travels in straight lines through a medium; however, the speed at which it travels is different, and depends on the density of the medium. When a ray of light passes from air into a solid such as glass, the change in velocity causes the ray to bend, its angle of entry moving towards the normal. As the ray emerges from the glass back into the air once more its velocity increases, and it bends back towards its original angle of incidence. This is illustrated in Figure 4.8(a).

The effect on the path of a ray passing through a number of differing media is shown in Figure 4.8(b). Because the refractive index of water is less than it is for glass the angles of refraction are different. However because, in this example, the ray is entering and emerging through the same medium (air), the angles of incidence and emergence (θ) are the same, although the paths are displaced.

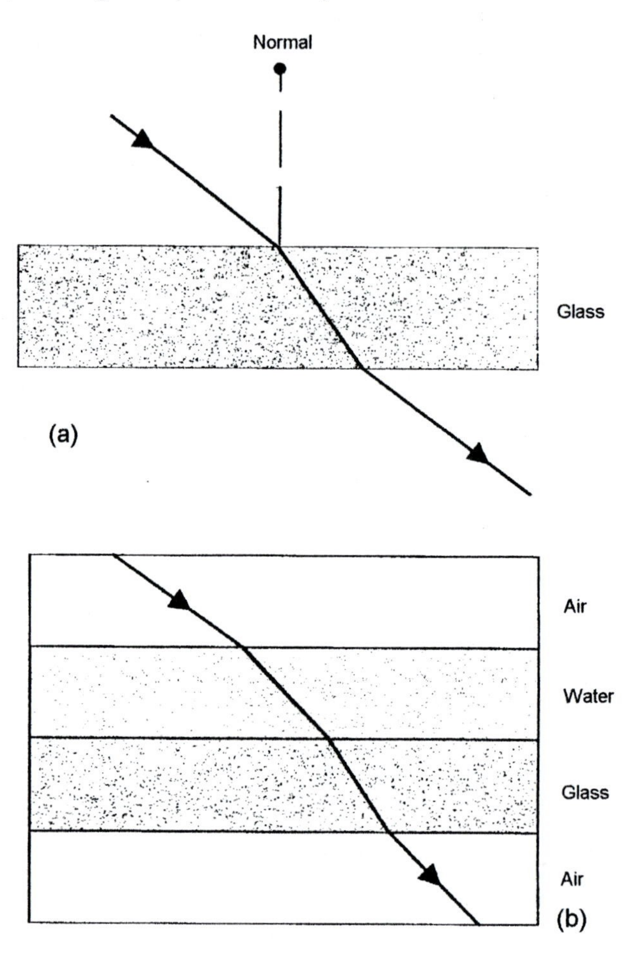

Figure 4.8
(a) Effect of a solid on the path of a light ray.
(b) Effect of differing media on a light ray

The amount by which a ray is refracted depends upon the change of velocity, which in turn is governed by the density of the solid. Each solid has its own *refractive index* which is found by dividing the velocity of light in a vacuum by the velocity of light within that solid. However, it is not only the refractive index which governs the angle of refraction. The frequency (i.e. the colour) of the light ray also has a bearing. Experimentation proves that for any given solid, the blue end of the light spectrum will always refract more than the red.

This effect is demonstrated in a *prism*, which is a flat, transparent medium with polished surfaces. When a ray of white light is passed through a prism with its surfaces cut to an angle of 60°, the differing refractive index for each frequency (colour) of light results in the lower frequencies following a shorter path than the higher frequencies. This is illustrated in Figure 4.9. The result is that when the original ray of white light emerges from the prism it appears as a rainbow. This effect is known as diffraction or dispersion.

Figure 4.9
A prism separates the wavelengths of light because they are refracted by different amounts

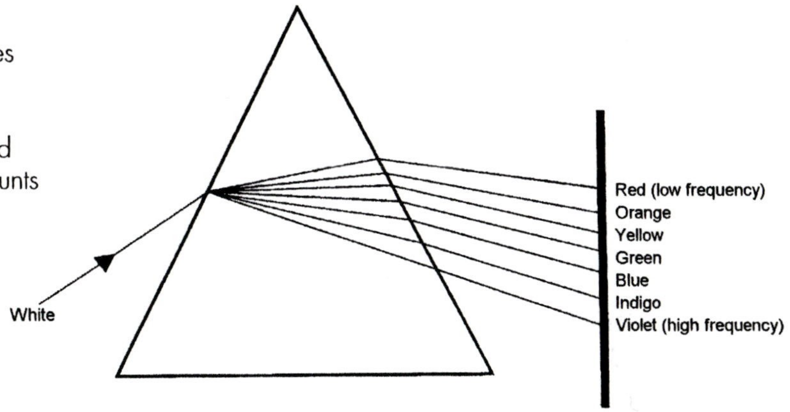

An optical lens is a device which makes use of the refractive effect on light paths. There are two types of lens: *convex* and *concave*.

A simple convex lens is shown in Figure 4.10(a). The light rays entering the lens are refracted; however, because the lens surface is curved, the angle of emergence at each point on the lens is different. If the lens is ground accurately then all of the rays of light will converge at a single point somewhere behind the lens. This point is called the *focal point*.

The concave lens is shown in Figure 4.10(b). This is known as a diverging lens because the light rays are bent outwards. In this case the focal point is said to be at the point on the entry side of the lens where the light appears to have originated.

One problem encountered with a simple lens is *chromatic aberration*. Remember that different colours of the spectrum have a different refraction index, so for the lens in Figure 4.10(a) the red end of the spectrum will not be bent as far inwards as the blue end. This means that there are a number of focal points, i.e. one for each wavelength in the light spectrum, resulting in an image that appears to have a number of coloured halos around it. In effect, the lens is behaving to some degree like a prism. The problem of chromatic aberration can be overcome by using a lens assembly comprising a series of converging and diverging lenses each made from glass with differing refractive index.

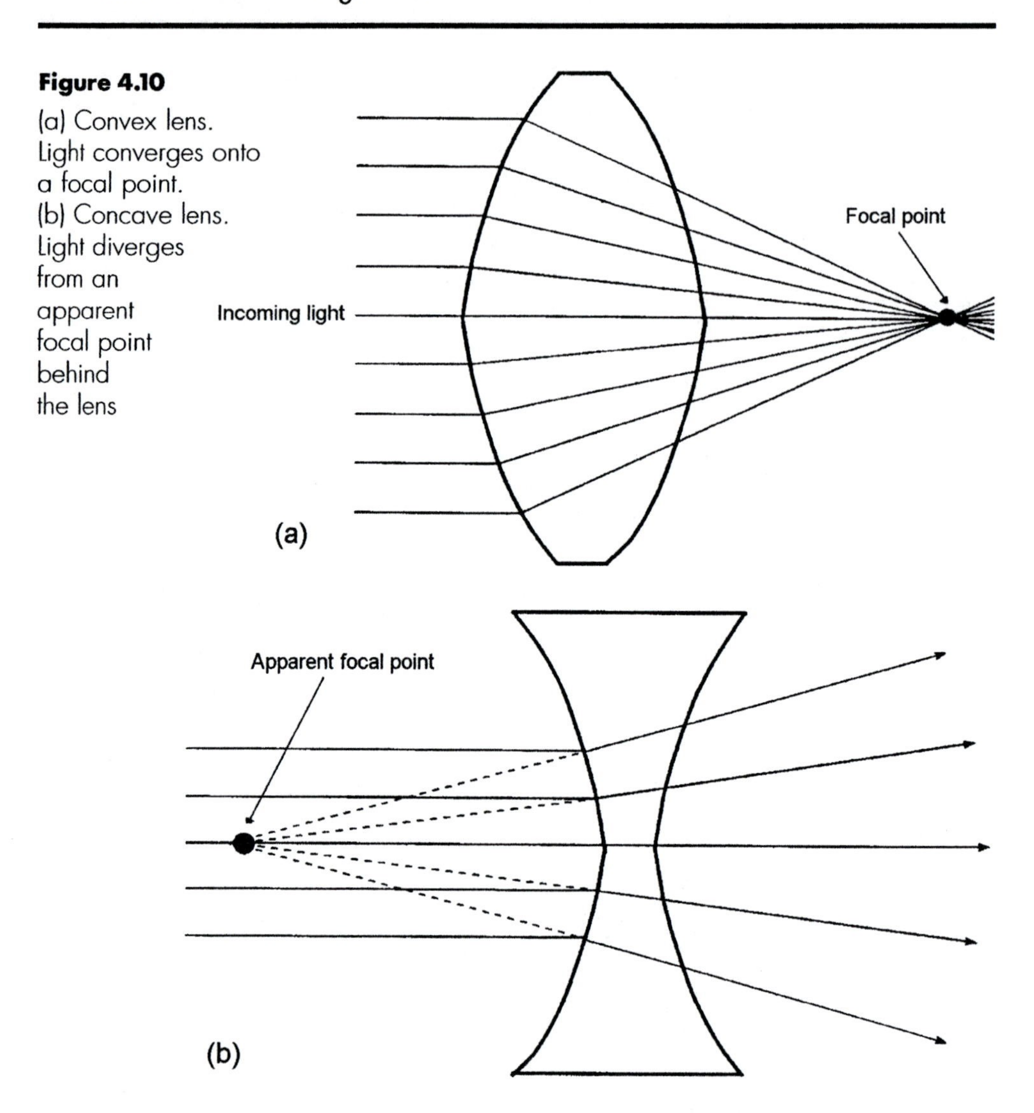

Figure 4.10
(a) Convex lens. Light converges onto a focal point.
(b) Concave lens. Light diverges from an apparent focal point behind the lens

Sound

Sound is 'heard' because a diaphragm in the ear vibrates. These vibrations are converted into electrical impulses which are fed to the brain. So it follows that anything which can cause this diaphragm to vibrate (or oscillate) can produce sound. Notice that the diaphragm must be made to oscillate, so the pressure causing the oscillation must take the form of a wave.

Just as the eye transforms electromagnetic waves into electrical impulses, the ear transforms mechanical waves into electrical impulses which, like electromagnetic waves, have the properties of frequency and wavelength.

Figure 4.11 shows a simple loudspeaker diaphragm. If the diaphragm is oscillating outwards (direction A) and inwards (direction B), then a series of pressure waves will move away from the diaphragm. As the diaphragm moves outwards the air immediately in front becomes compressed. This increase in pressure is then transferred to the air further away, hence a wave of high pressure moves away from the speaker. When the diaphragm moves inwards the air becomes rarefied (decrease in pressure), resulting in a wave of rarefied air moving away from the speaker.

Continuous oscillation of the speaker diaphragm results in a series of acoustic waves, the frequency being equal to the frequency of movement of the diaphragm.

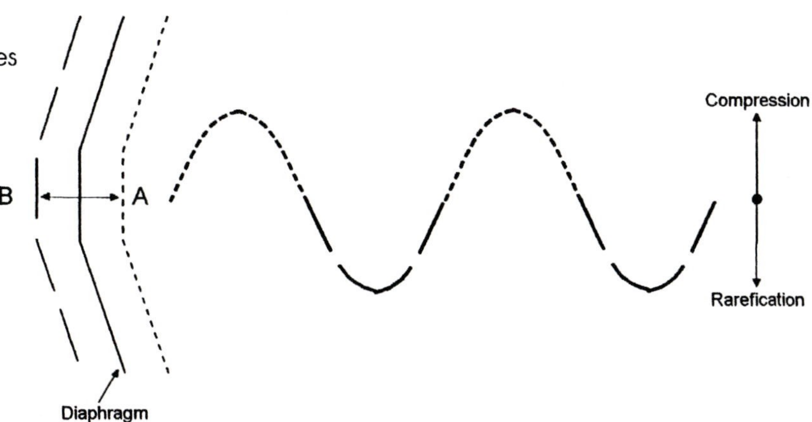

Figure 4.11
Acoustic waves produced by a moving diaphragm

Of course, being a mechanical wave, sound does not travel only through air. Any solid will oscillate and will therefore carry sound waves. The denser the solid, the better the wave propagation.

Sound propagates outwards in all directions from the source, assuming that it is permitted to do so. This is illustrated in Figure 4.12.

The velocity of propagation of sound varies depending upon the medium through which it is travelling. In dry air at a temperature of 0°C sound travels at a speed of 330 m/sec, but if the temperature is raised then the speed of propagation increases.

Because sound travels as a wave, then we can apply the same wave calculations as we do for light. So:

$$f = \frac{v}{\lambda}$$

where f = frequency, v = velocity of propagation (330 m/sec) and λ = wavelength.

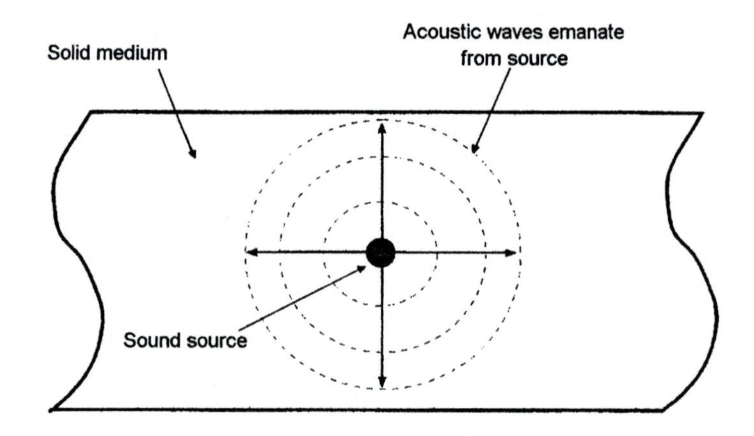

Figure 4.12 Acoustic waves travelling through a solid medium

Example

Calculate the wavelength of an acoustic wave that has a frequency of 1 kHz.

Solution

From the above equation, $\lambda = v \div f$

$$\therefore \lambda = \frac{330}{1000} = 0.33 \text{ m/sec}$$

5

Electricity

In order to understand electrical current, it is necessary to examine the atomic structure of the materials known as electrical conductors.

All matter is formed from molecules, and a molecule is the smallest particle of any substance that still retains the physical characteristics of that substance. A molecule can be made up from a compound of elements, or just one single element. Each of the 103 different elements, e.g. hydrogen, carbon, copper, uranium, etc., represents a different atom.

An atom consists of a number of particles. At its centre is a nucleus around which orbit a number of negatively charged particles known as electrons. Each of the 103 different atoms has a different number of orbiting electrons. The simplest atom (or element), helium, has just one orbiting electron (Figure 5.1(a)); the next, hydrogen, has two electrons, etc. Inside the nucleus of the atom are, among other particles that do not concern us at the moment, positively charged particles called protons. Each atom must have an equal number of protons and electrons.

The electron orbits (known as shells) are well defined, and for every atom there will be no more than two electrons in the first shell, eight in the second shell, eighteen in the third, thirty-two in the fourth, and so on up to seven shells. Unless forced to do so, no electron will remain in a higher orbit until all of the locations in the lower orbits have been filled. The electrons in the outer shell are known as valence electrons, unless the shell is complete where the atom is said to have no valence electrons. It is the number of valence electrons which determine whether an element is going to be an electrical conductor or an insulator.

Conductors and insulators

Figure 5.1(b) shows a representation of a copper atom. A single atom of copper contains 29 electron/proton pairs, distributed in four shells, and has just one valence electron in the outer shell. Where the outer shell is largely incomplete there is no strong cohesive bond between the valence electrons and the nucleus, and it therefore requires very little electrical force to make the valence electrons break out of their shell and move freely within the molecular structure of the element. This free movement of electrons makes it possible for the material to conduct electricity. Elements with three or less valence electrons are conductors of electricity.

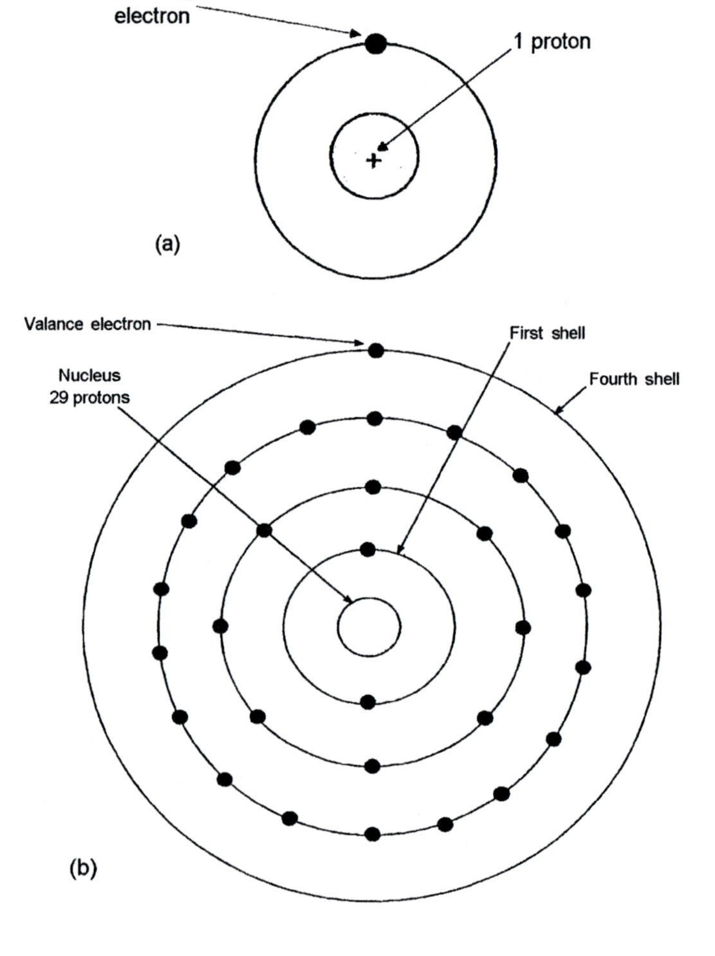

Figure 5.1
(a) Representation of a helium atom. (b) Representation of a copper atom

An atom with more than four valence electrons is far more robust, and unless an extremely high potential is applied, the atoms will remain stable and the material will be an electrical insulator.

Both conductors and insulators play an important role in electronics. Table 5.1 lists some common conductor and insulating materials.

Table 5.1

Conductors	*Insulators*
All metals	Ceramic
Silver (best)	Glass
Copper (most common)	Mica
Aluminium	Plastics
Carbon	Pure water
Acid; alkaline or salt solutions	Dry air

There are some elements which under normal conditions are electrical insulators, but if other elements are added to their molecular structure they will conduct electricity at normal temperatures. Such materials are known as *semiconductors*. The most common examples of these are silicon and germanium, both of which have four valence electrons.

Electrical current

If, as shown in Figure 5.2, positive and negative charges are placed across a conductor, the negatively charged valence electrons will be attracted from their shells towards the positive charge. This will leave the atoms in an unbalanced state as they will now have more protons in their nucleus than they have orbiting electrons. To stabilise this condition, electrons will be attracted from the applied negative charge and will move into the vacated shell locations. There will now be a continuous movement of electrons through the conductor as they move from the negative charge towards the positive charge, leapfrogging from atom to atom as they progress. This movement of electrons through a conductor is called an *electric current*.

Figure 5.2
Electron movement through a conductor

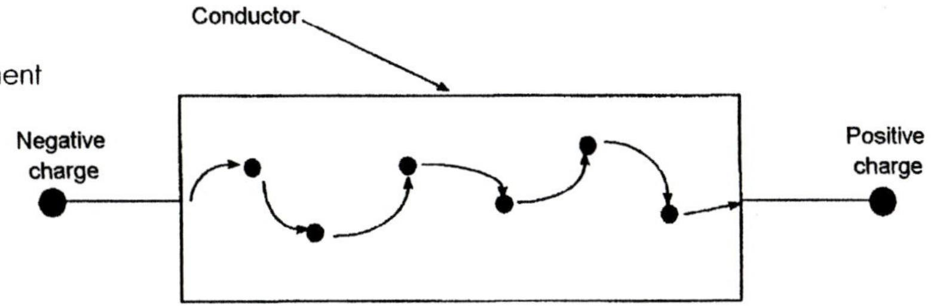

The unit of measurement of current is the *ampere,* commonly abbreviated to 'amp' or 'A'. Just as we can see that the unit of one gallon of water is derived from a specific amount of water, so the ampere is defined by a specific number of electrons passing a given point in one second. The scientific reading is taken by measuring the force exerted between a permanent magnet and a current carrying conductor. When that force equals a specified value, then the total number of electrons passing a point in one second will be 6.289×10^{18}. This unit of electrical charge is known as the *coulomb,* so in other words, when one coulomb of charge passes a point in a conductor in one second, the value of current will be one ampere (1 A).

Although electricians who deal with high power electrical equipment may talk in terms of hundreds of amps, the majority of circuits that an electronics engineer will encounter draw only a fraction of one amp. For this reason the units of milliamps (thousanths of one amp), or microamps (millionths of one amp) are frequently used.

Figure 5.2 shows that the current is flowing from the negative charge to the positive charge. This direction of current is known as *electron flow* and is the true direction of movement of electrical current. However, it is still common practice to refer to *conventional current* which is said to flow from positive to negative. Conventional current dates back many years when the early pioneers of electricity believed that current actually flowed in this direction.

When a current flows around a circuit there will be manifest at least one of three effects within that circuit: heating, magnetic, or chemical.

The heating effect is a result of the electrons losing energy as they move through the circuit, and one way in which this energy is manifest is in the form of heat. This will be examined later in the chapter when considering electrical power.

When an electron moves through a circuit, a minute magnetic field will exist around that electron. Since an electrical current is made up from many millions of electrons, the magnetic fields from all of the electrons merge to form one large magnetic field around the conductor through which the current is flowing. This effect finds many applications in electrical and electronic circuits and is employed in loudspeakers, microphones, motors, generators, and relays.

The chemical effect of current is best illustrated by the electric cell (a cell is one part of a battery). When a current is passed through a liquid containing a solution of chemicals, the current breaks the compound down into its component parts, one part gaining an excess of electrons giving it a negative charge, the other having a deficiency of electrons giving it a positive charge. Cells and batteries will be discussed in more detail in Chapter 7.

The chemical effect of an electric current finds other applications in industry, for example the electroplating of metals.

Electromotive force

It has been stated that the application of a charge across a conductor causes a flow of electrical current, therefore this charge must be exerting a force on the electrons. This force is called the *electromotive force (e.m.f.)*. If the e.m.f. is increased, the current will increase proportionally. If the e.m.f. is removed, current flow ceases. If the polarity of the e.m.f. is reversed, then the direction of current flow will reverse.

By virtue of the fact that electrons have been made to move, energy has been used and work has been done. If the rate of doing work is one watt when the current is one ampere, then the e.m.f. applied is said to be one volt (1 V).

An e.m.f. can be produced by either chemical (battery), magnetic (generator), or thermal means. For example, in a battery cell, as a result of electrochemical effect, electrons move from one plate in the cell to the other. This results in a surplus of electrons on one plate giving it a negative charge, and a deficiency of electrons on the other plate resulting in a positive charge.

Just because an e.m.f. is present does not mean that current must flow. There must be a conductive circuit connected across the e.m.f. to provide a path through which the electrons can move. For example, a battery has an e.m.f. across its terminals, but current will not flow until the battery is connected into a circuit.

Electrical resistance

So far we have seen how by applying an e.m.f. to a conductor we can cause electrons to move through that conductor. However, we have assumed that the electrons are able to move unimpeded, but in reality this is not so. In any conductor there is a random movement of electrons because weakly bonded valence electrons constantly break from their atom, move through the material and join to another atom that has also lost a valence electron. This movement of electrons impedes the flow of current through the conductor. This opposition to current flow is termed *resistance*. The unit of electrical resistance is the *ohm*, and we use the Greek symbol Ω.

Because different types of conductor, e.g. copper, aluminium, gold, etc., do not have the same number of valance electrons, the degree of electron movement within conductors varies, which means that they each have a different resistance.

For any given conductor, if the cross-sectional area is doubled, the area through which electrons pass is doubled, so the resistance is halved. Conversely, doubling the length of the conductor will double the resistance. This relationship is expressed in the formula:

$$R \propto \frac{L}{A}$$

where the symbol $\propto$ means 'is proportional to'. This relationship is illustrated in Figure 5.3.

Figure 5.3 Effect of conductor length and area on resistance

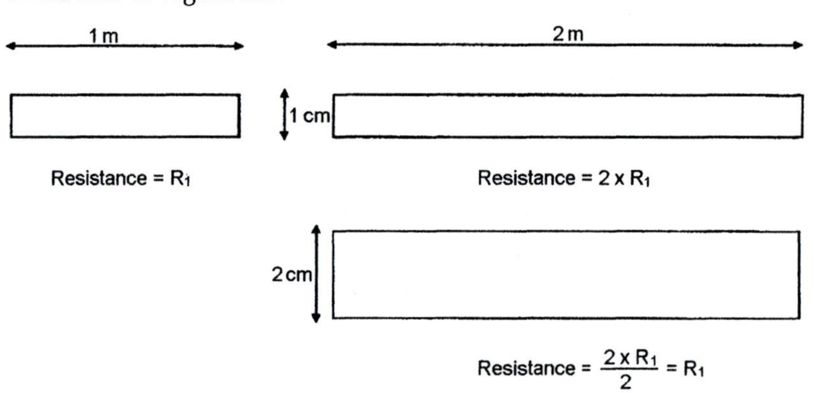

In order to calculate the resistance of a length of conductor we need to know the precise resistance of the material. As we have seen, the resistance varies for each type of conductor, and the problem is made more difficult because the movement of free electrons within the material varies with temperature. Thus, the resistance of every material used in the electronics industry has been carefully measured and the values are given along with the temperature at which they were taken. The resistance for each material is known as its *specific resistance*, and is expressed in units of ohms per metre (Ω/m). The symbol for specific resistance is the Greek letter ρ, pronounced 'rho'. Samples of specific resistance for some common materials are given in Table 5.2.

The resistance of any conductor can thus be found using the expression:

$$R = \rho \times \frac{L}{A}$$

where ρ = specific resistance of conductor, L = length of conductor and A = cross-sectional area of conductor (in m^2).

Table 5.2 Approximate values of specific resistance for some common materials

Material	*Resistivity* (at 20°C (μΩ/m))
Aluminium	0.028
Silver	0.016
Copper	0.017
Iron	0.098
Glass	1×10^{10}
P.T.F.E.	1×10^{18}

Example

Determine the resistance of 500 m of copper wire with a diameter of 0.5 mm.

Solution

From Table 5.2 we see that the specific resistance of copper is 0.017 μΩ/m. The cross-sectional area will be $(\pi \times r^2) = 3.142 \times 0.00025\ \text{m}^2 = 196 \times 10^{-9}$. Thus:

$$R = \frac{0.017 \times 10^{-6} \times 500}{196 \times 10^{-9}} = 43.4\,\Omega$$

One of the most common components used in electronic circuits are resistors. These are components which are manufactured to have a certain resistance and, as we shall see later on, are employed to deliberately limit the current flow in a part of a circuit. Two British Standard symbols for resistors are shown in Figure 5.4, but the one most commonly employed is the rectangular box design.

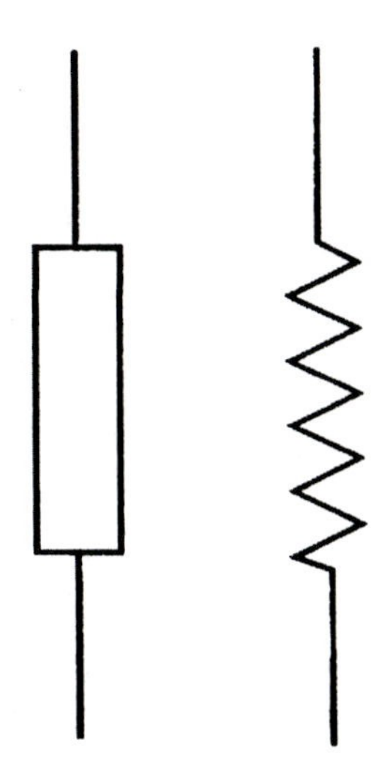

Figure 5.4
British Standard symbols for resistors

Ohm's law

Ohm's law brings together the three quantities, voltage, current, and resistance, and defines the relationship between them.

Just to recap, when an e.m.f. is applied to a circuit a current will flow. Increasing the e.m.f. results in a proportional increase in current. Resistance in the circuit opposes the flow of current, and an increase in the circuit resistance results in a decrease in current. So, *the current in a circuit is directly proportional to the applied voltage, and is inversely proportional to the circuit resistance.*

Ohm's law expresses this relationship as:

$$E = I \times R$$

where E is the applied e.m.f. (frequently shown as V), I is the current flowing in the circuit and R is the circuit resistance.

The expression can be transposed to find:

$$I = \frac{E}{R} \quad \text{and} \quad R = \frac{E}{I}$$

These equations are fundamental to understanding electrical and electronic circuits, and you will continue to meet them throughout your work in electronics.

To demonstrate the application of the Ohm's law equation, three simple circuits are given in Figure 5.5 with accompanying solutions.

Example

A cooker element passes a current of 3 A when 230 V is applied across it. What is the resistance of the element?

Solution

$$R = \frac{230}{3} = 76.7\ \Omega$$

Example

A 9 V battery has a load of 1 kΩ across it. What is the current taken from the battery?

Solution

$$I = \frac{9}{1000} = 0.009 \text{ A or } 9 \text{ mA}$$

Potential difference

We know that application of an e.m.f. causes current to flow. This means that an amount of energy has been given to the electrons, and this energy is measured in volts.

Figure 5.5
Use of Ohm's law to determine quantities in a circuit

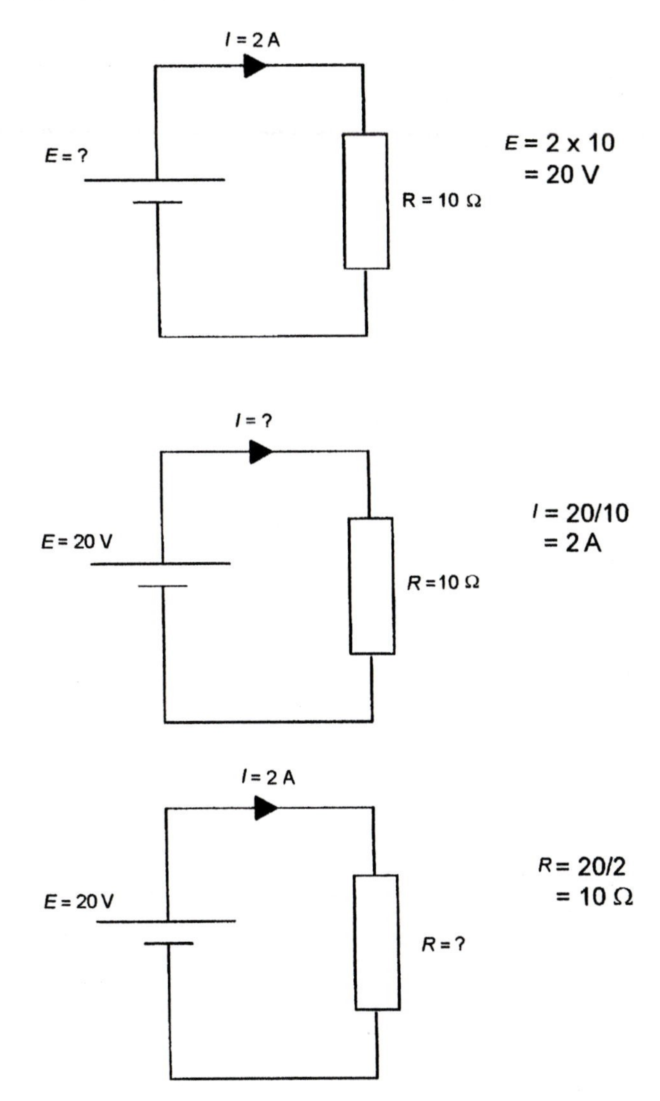

As the electrons move around the circuit and meet with the circuit resistance, they lose some of their energy. This energy loss is called the *potential difference* (*p.d.*), and, because the energy given to the electrons is measured in volts, the p.d. is measured in volts. To distinguish between an e.m.f. and a p.d. voltage, we use the symbol '*v*' for potential difference.

By definition, when a current of 1 A dissipates a power of 1 watt, the potential difference is said to be 1 volt.

Remember, an e.m.f. causes a current to flow; a p.d. is caused by the flow of current.

We shall look at potential difference in more detail later on.

The joule

This is the SI unit of energy and by definition, when an object is moved through a distance of 1 metre with an applied force of 1 newton, the work done is said to be 1 joule (1 J). Although this is a mechanical definition, we shall see in a moment how it translates into a unit of electrical energy.

Power

In terms of physics, power is the *rate* of doing work or dissipating energy, i.e. when a certain amount of work, expressed in joules, is done in a certain amount of time, expressed in seconds, the rate of doing work can be expressed in joules per second. The rate of 1 joule per second is the unit of power and is termed the watt (W). Thus, 1 J/sec = 1 W.

In an electrical circuit the work 'done' is the losing of energy by the electrons, and in electrical terms 1 J/sec can be equated to a power of 1 watt (1 W) when 1 A flows causing a p.d. of 1 V.

Electrical devices and components all have a power rating which is determined by the maximum amount of power they can dissipate or convert. For example, a 100 W light bulb dissipates 100 joules of energy per second. The relationship between mechanical and electrical energy can be seen when we consider that a 100% efficient 100 W electric motor would convert 100 joules per second of energy into motion.

We have seen that when current flows through a resistance, a voltage is developed. Therefore, we can say that:

Power (P) = $V \times I$

This means that if we know the current and voltage (p.d.) in a circuit or component, we can calculate the amount of power that is going to be dissipated in that circuit or component. This is important when we have to select the size or rating of a component because electrical power is dissipated mainly in the form of heat, and if we select a component that is not large enough to dissipate this heat then early failure will occur, with the possibility of a fire.

There will be instances where we do not know both the current and the voltage, but if we know the resistance of the circuit in which we wish to calculate the power dissipation we can apply Ohm's law to derive the following expressions.

(A) if both I and R are known:
Substituting $I \times R$ for V we have

Power = $(I \times R)I = I^2 \times R$

(B) if both V and R are known:
Substituting V/R for I we have

$$\text{Power} = \frac{V}{R} \times V = \frac{V^2}{R}$$

Any one of these three equations may be used to calculate power.

For a simple application of these equations, look again at the three circuits in Figure 5.5. In each case we shall be using the two known quantities to calculate the power dissipated in the resistor.

Circuit A: I and R are known thus, using $P = I^2 \times R$:

$$P = 2^2 \times 10 = 40 \text{ W}$$

Circuit B: Both V and R are known (in this case the p.d. across the resistor is equal to the e.m.f.), thus, using $P = V^2/R$:

$$P = \frac{20^2}{10} = 40 \text{ W}$$

Circuit C: Both I and V are known, thus, using $P = V \times I$:

$$P = 20 \times 2 = 40 \text{ W}$$

One watt is a considerable amount of power, and it is therefore very common to use a smaller unit, i.e. milliwatts (mW).

Temperature coefficient

All materials, both conductors and insulators, exhibit a change of resistance with temperature. In the case of most conductors the resistance increases with a rise in temperature. With semiconductors and most insulators the resistance decreases with a rise in temperature. This change in resistance does not follow a linear law, but over a small temperature change, it may be considered to do so.

The change in resistance is measured by the *temperature coefficient* which is defined as the fractional increase or decrease of resistance per degree change in temperature. For example, a resistor would be quoted as having a temperature coefficient of something in the order of 300 ppm/°C (ppm = parts per million), which means that for every 1°C change in temperature its resistance will change by 300 Ω for every 1 000 000 Ω of resistance, or 0.3 Ω for every 1000 Ω.

Example
A 10 kΩ resistor is quoted as having a temperature coefficient of 300 ppm/°C. If its resistance measures 10 kΩ at 20°C, what will its resistance be at 100°C?

Solution
Change in temperature = 80°. Resistance change = 0.3 Ω/1k. Therefore:

$$\text{Change of resistance} = \frac{0.3}{1000} \times 10\,000 \times 80 = 240\ \Omega$$

New resistance value = 10 kΩ + 240 Ω = 10.24 kΩ.

Components that exhibit an increase in resistance with rise in temperature are said to have a *positive temperature coefficient (p.t.c.) of resistance*. Conversely, those that reduce their resistance with a rise in temperature have a *negative temperature coefficient (n.t.c.)*.

There are special resistors called thermistors which are designed to make use of the effect of temperature coefficient. We shall look at these in Chapter 6.

6

Resistors and resistive circuits

Resistors are used primarily to limit the current in, and apply the correct voltage to, a part of a circuit and are one of the main building blocks in electronic circuits. There is a wide range of resistors available covering not only different resistive values but also tolerance values, power ratings, and construction. The most common types of construction are carbon composition, carbon film, metal film, and wirewound. Surface mount devices (SMD) are also used extensively in production of all types of electronic equipment, but we shall look at these separately in Chapter 11.

Although they are relatively cheap to manufacture and thus purchase, carbon composition resistors are rarely used in modern production; because of the way in which they are manufactured they do not have a very high tolerance. We shall look at resistor tolerance in more detail in a moment.

Carbon and metal film resistors are the best devices for general-purpose applications because they are obtainable in high tolerances and are relatively inexpensive.

Resistor construction

As shown in Figure 6.1, film resistors are constructed by wrapping a carbon or resistive metal film around an insulated tube. A spiral cut is then made around the resistor. The longer the spiral cut, the longer the resulting current path and hence the greater the resistance (remember that the resistance is proportional to $\rho \times l \div a$).

Film resistors are available in power ratings ranging between 1/8 watt and 2 watt. Powers above 2 W are available; however, they are not favoured because they are rather large and expensive.

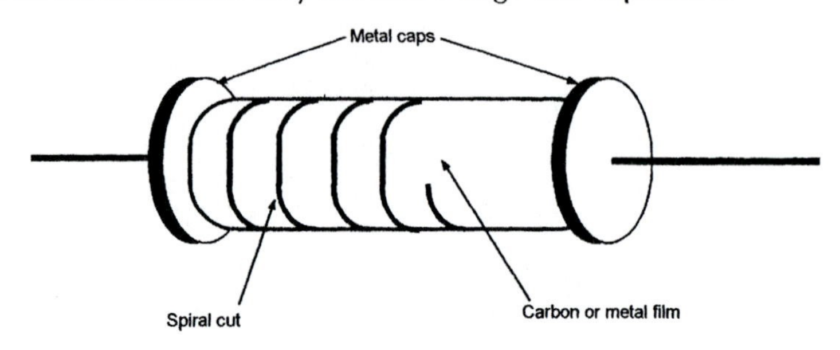

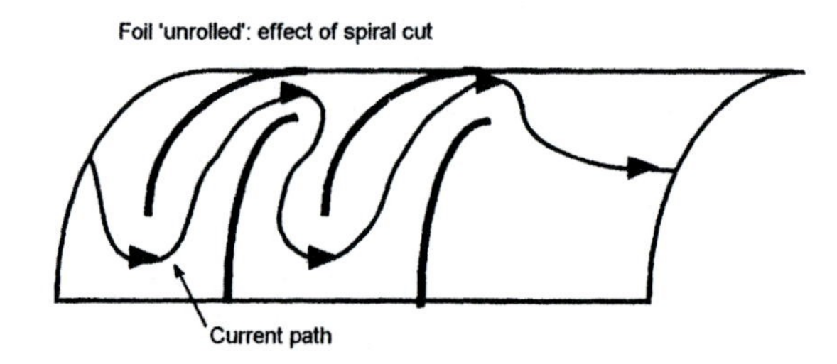

Figure 6.1 Carbon or metal film resistor construction

For powers above 2 W, *wirewound resistors* are usually employed. These are constructed by wrapping a resistive wire around a ceramic former as shown in Figure 6.2. The value of resistor is determined by the resistance of the wire used, and the length of wire. For insulation purposes the resistor may be coated in ceramic, or the former may be encapsulated inside a ceramic case.

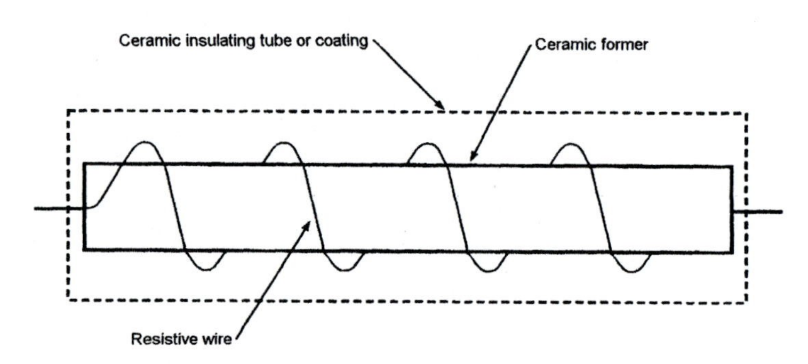

Figure 6.2 Wirewound resistor construction

To increase the power rating heavier duty wire and a larger ceramic former is used to help dissipate the heat. The maximum power available at a reasonable size is 17 W, although much higher operating temperatures are available with tubular construction to permit airflow, and heat sinks to give better heat dissipation. A typical wirewound resistor is shown in Figure 6.3.

Figure 6.3
Ceramic wirewound resistor

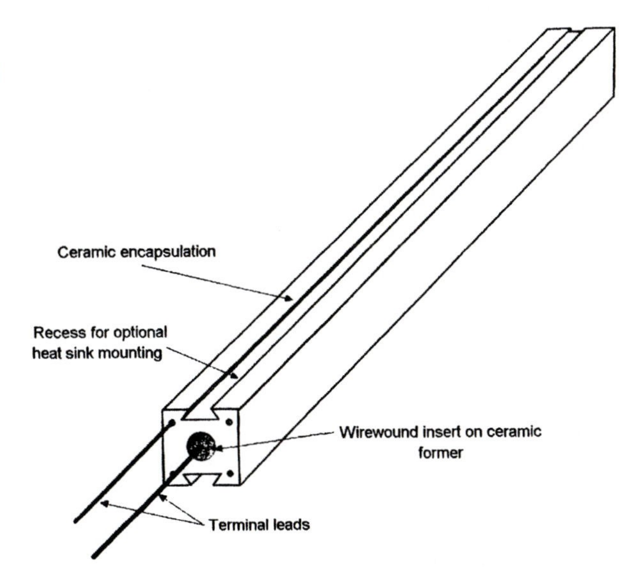

Resistor values

It is not practical for manufacturers to offer a resistor of every possible value as the range would have to include literally millions of resistors! Instead a selection of values, known as *preferred values*, are offered.

In addition to the value of a resistor, we must also take into account the *tolerance*. Mass production and manufacturing to a price means that a resistor that has been labelled, say, 1 kΩ could in reality have a value that is slightly higher or lower. This variation is accounted for by giving the resistor a tolerance value. So, for example, a 1 kΩ resistor with a 10% tolerance could have an actual value of 1 kΩ + /− 100 Ω. In other words, it may have a value of anything between 900 Ω and 1.1 kΩ.

During production the values of the resistors are measured, and they are placed into batches relating to their tolerance. The price of the resistor increases as the tolerance value improves.

Although tolerance must be taken into consideration by circuit designers, many circuits will happily accept a small tolerance in component values. Certain circuits, such as those used in measuring equipment, e.g. test meters and oscilloscopes, require very high tolerance components to ensure accuracy. For such equipment, resistors with a tolerance as low as 0.1% are available; however, these components are very expensive which is one reason for the high cost of such equipment.

Table 6.1 shows the range of preferred values for 5%, 10%, and 20% tolerance resistors. The figures were chosen so that any resistor that is manufactured will fall within the tolerance range of one of the preferred values, therefore no component has to be discarded on account of its value alone. Each numerical value is applied across the range of resistor values; for example, the number 47 will be found as 4.7, 47, 470, 4.7k, 47k, etc.

Table 6.1

5%	10	11	12	13	15	16	18	20	22	24	27	30	33	36	39	43	47	51	56	62	68	75	82	91
10%	10		12		15		18		22		27		33		39		47		56		68		82	
20%	10				15				22				33				47				68			

20% tolerance resistors are rarely used in modern equipment, and the majority of modern circuits employ 5%, and occasionally 10%, tolerance resistors.

Resistor coding

In the case of larger resistors such as wirewound types, the value and tolerance (and in some cases the power rating) can be printed numerically on the side. However, this is more difficult when dealing with small carbon and metal types for a number of reasons. The writing would have to be very small making it in some cases illegible, the printing process would add to the cost of the resistors, and the print would be liable to scratch off when the resistors are stored.

For these reasons a colour code was devised (although at the time that it was devised the primary reason was cost, as resistors were so large that there was more than adequate space to print onto them!). The resistor colour code is given in Table 6.2, and it is highly recommended that an engineer makes it an early priority to memorise these ten colours and numbers, as it is something that they will use frequently.

Table 6.2

Black	=	0	Green	=	5
Brown	=	1	Blue	=	6
Red	=	2	Violet	=	7
Orange	=	3	Grey	=	8
Yellow	=	4	White	=	9

There are two common applications of the code, four band and five band, and both are quite simple to use. We shall look first of all at the four band application.

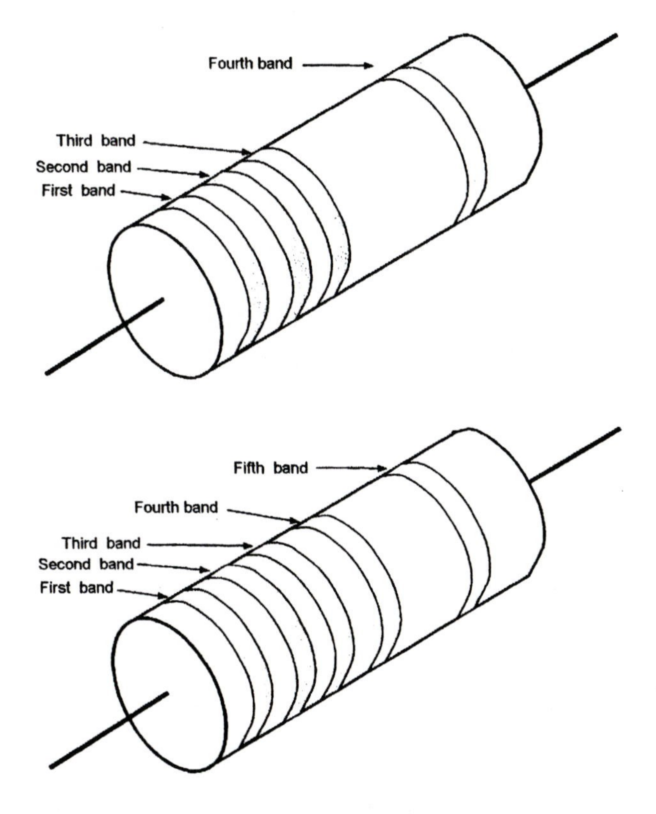

Figure 6.4
Four and five band coded carbon resistors

Looking at Figure 6.4 we see that the colours are painted towards one end of the resistor. This end must be held at the left-hand side. *The first colour then becomes the first number, the second colour gives the second number, and the third colour* (called the multiplier) *denotes the number of zeros (0s) in the value.* For example, a resistor having the first three colours brown, black, red would have a value of:

from Table 6.2   brown = 1   black = 0   red = 2

therefore we have 1, 0, and two 0s giving us 1000 Ω, or 1 kΩ.

The fourth band tells us the percentage tolerance of the resistor. The colour code for the tolerance band is given in Table 6.3. Where there is no tolerance band on the resistor, this indicates a tolerance of 20%.

Example
Find the value of a resistor with the code green, blue, red, silver.

Solution

Green = 5, blue = 6, and red = two 0s.
Therefore we have 5600 Ω or 5.6 kΩ, with a 10% tolerance giving 5.6 kΩ + /− 560 Ω.

Example

Find the value of a resistor with the code brown, black, black, gold.

Solution

Brown = 1, black = 0, and black = no 0s.

Therefore we have 10 Ω, with a 5% tolerance giving 10 Ω + /− 0.5 Ω.

In the first example we came to a value of 5.6 kΩ. To simplify circuit diagrams this expression is frequently shortened to 5k6, where the position of the decimal point is denoted by the position of the 'k'. The same applies to the 'M' where the value is in megohms, and where the value is simply in ohms, an 'R' is used to indicate the decimal position; e.g. 3.3 Ω would be written 3R3, The 'Ω' sign frequently being omitted to further shorten the expression.

The second example reveals a problem with the code that must be dealt with. Using the code as we have done so far means that we cannot indicate a value of less than 10 Ω. Try and denote 9 Ω, you will find that it is not possible; unless you leave off the third band, but then you would not know if it had simply been scratched off during storage!

Values below 10 Ω are denoted by using a multiplier band with a value of less than 1 (see Table 6.3). Where the third band is gold, the value is multiplied by 0.1 (or divided by 10). Where it is silver, the value is multiplied by 0.01 (or divided by 100).

Table 6.3

Colour	*Digit*	*Multiplier*	*Tolerance*
Black	0	× 1	
Brown	1	× 10	1%
Red	2	× 100	2%
Orange	3	× 1k	
Yellow	4	× 10k	
Green	5	× 100k	0.5%
Blue	6	× 1M	0.25%
Violet	7	× 10M	0.1%
Grey	8	not used	
White	9	not used	
Silver		× 0.01 (100)	10%
Gold		× 0.1 (10)	5%

Example

Find the value of a resistor with the code orange, orange, gold, gold.

Solution

Orange = 3, and gold = $\times 0.1$.

Therefore we have $33 \times 0.1 = 3.3\,\Omega$ with a 5% tolerance giving $3.3\,\Omega +/- 0.17\,\Omega$.

Application of the five band code is very similar to the four band. In this case the *first three colours denote the first three numbers, and the fourth band is the multiplier*. The fifth band is still the tolerance band. This code is used for resistors that have a more precise value. For example, we could have brown, brown, red, brown, gold. This gives us a value of 1120 Ω, or 1k12 Ω, with a 5% tolerance.

In some cases a sixth band is in evidence. This is used to denote the temperature coefficient, but such resistors are generally only found in specialised electronic equipment.

The power rating of carbon and metal film resistors is not indicated on the device, and it is sometimes difficult to decide what the rating of a particular resistor actually is, although experience helps to a large degree. When replacing a defective resistor, the care must be taken to ensure that the power rating of the new device is at least as high as the one you are removing, otherwise early failure may occur, with the possibility of an electrical fire.

Resistors generally either go open circuit, or increase in value when they fail, although very high values do sometimes fall in value. When testing a resistor, it should be removed from the circuit and its value read on a meter; the method is shown in detail in Chapter 15. The reading on the meter should fall within the tolerance range indicated on the resistor.

Specially designed resistors, both film and wirewound types, are frequently employed as safety devices (fuses). These are discussed in more detail in Chapter 11.

Resistive circuits

In the last chapter we examined the Ohm's law relationship between voltage, current, and resistance. We shall now look at how this can be used to analyse electrical and electronic circuits.

Resistors in series

When two or more resistors are in *series* across an e.m.f., a current will pass through each one, meeting resistance as it progresses (Figure 6.5).

Thus, *for resistors in series, the total resistance is the sum of all the resistances*; i.e. for the circuit in Figure 6.5, $R_t = R_1 + R_2 + R_3$.

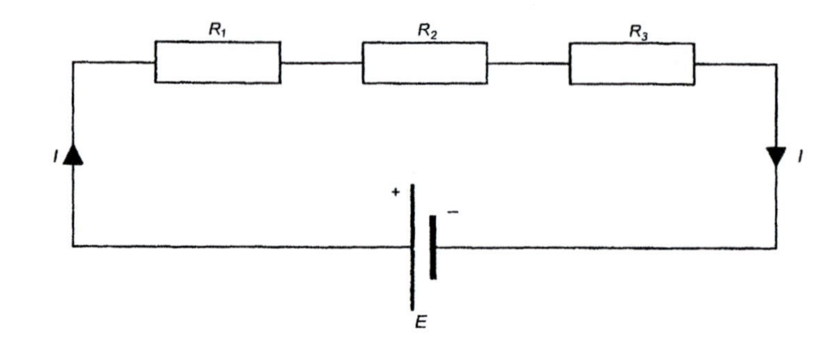

Figure 6.5
In a series circuit the p.d. across each resistor is different, but the current is common

The p.d. across each resistor depends upon the value of each resistor, but note that *the sum of the p.d.s must equal the applied e.m.f.*

Example

In Figure 6.6, the current is 1 A. Thus, using Ohm's law; from $V = I \times R$:

the p.d. across R_1 $(VR_1) = 1 \times 6 = 6$ V
and the p.d. across R_2 $(VR_2) = 1 \times 4 = 4$ V:

Notice that the sum of the p.d.s is equal to the e.m.f.

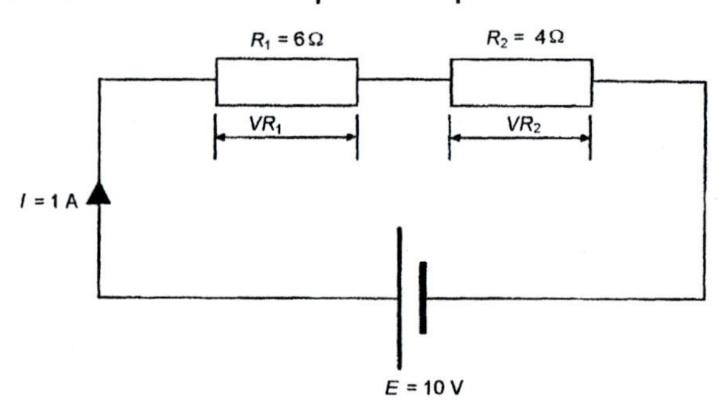

Figure 6.6

The following examples show how Ohm's law can be used to find different quantities in series circuits. Sometimes it may appear that we do not have enough information to resolve the problem, but if we apply our knowledge of the behaviour of electrical circuits we can very quickly formulate an answer.

Example

For the circuit in Figure 6.7, calculate the p.d. across each resistor.

Solution

First we must find I, which is equal to E/R_t.
$R_t = R_1 + R_2 + R_3 = 2 + 5 + 3 = 10\ \Omega$.

Thus $I = 20 \div 10 = 2$ A.
$VR_1 = I \times R_1 = 2 \times 2 = 4$ V.
$VR_2 = I \times R_2 = 2 \times 5 = 10$ V.
$VR_3 = I \times R_3 = 2 \times 3 = 6$ V.

Figure 6.7

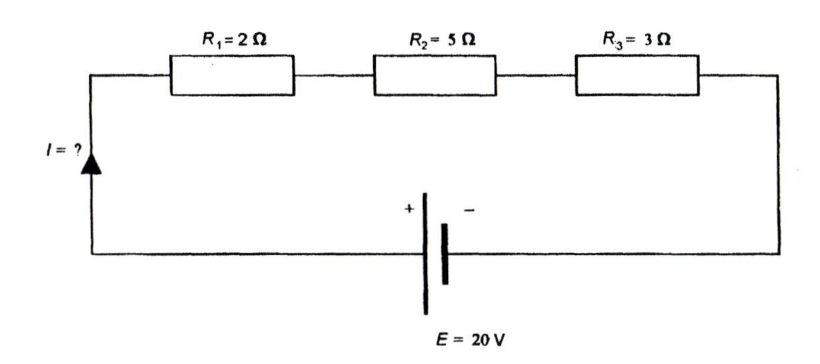

Example

For the circuit in Figure 6.8, calculate the applied e.m.f.

Solution

First find the p.d. across R_3 using $V = I \times R_3 = 2 \times 3 = 6$ V.
Because the applied e.m.f. must equal the sum of the p.d.s.
$E = VR_1 + VR_2 + VR_3 = 7\text{ V} + 2\text{ V} + 6\text{ V} = 15$ V.

Figure 6.8

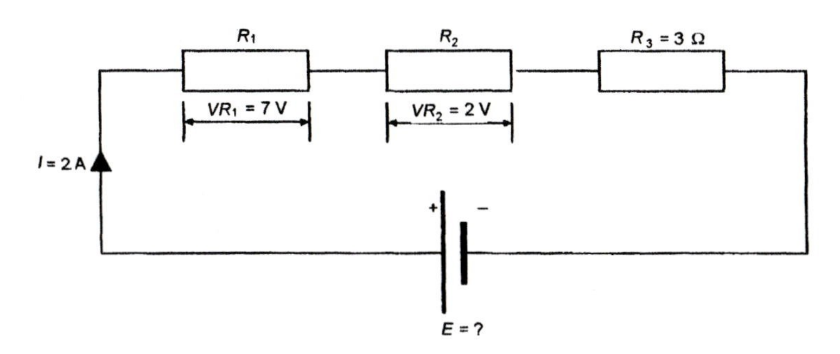

Resistors in parallel

When resistors are connected in parallel across an e.m.f., *the p.d. across each resistor will be the same, but the current flowing around the circuit will divide between the resistors.* If, as in Figure 6.9, the resistors are all of the same value, the current divides evenly.

No matter how many parallel circuits there are, the current will still divide between the circuits, and these portions of current will all merge once more before returning to the source e.m.f.

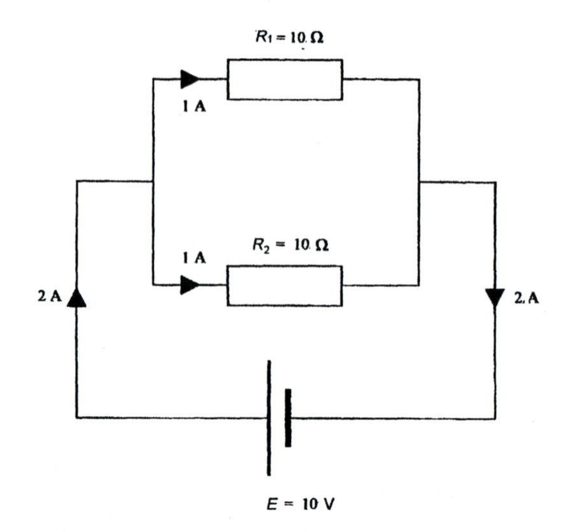

Figure 6.9
In a parallel circuit the current divides, but the p.d. across each resistor is the same

By connecting two resistors in parallel we effectively increase the cross-sectional area through which current can flow. This means that the total resistance has been reduced. For example, in Figure 6.9 two 10 Ω resistors connected in parallel will offer twice the area through which current can flow compared to a single resistor. Therefore the current will be double, which means that the total resistance must have halved.

When two resistors of differing values are connected in parallel, the current will divide proportionally; the higher proportion of current flowing through the lower value of resistor. This is illustrated in Figure 6.10.

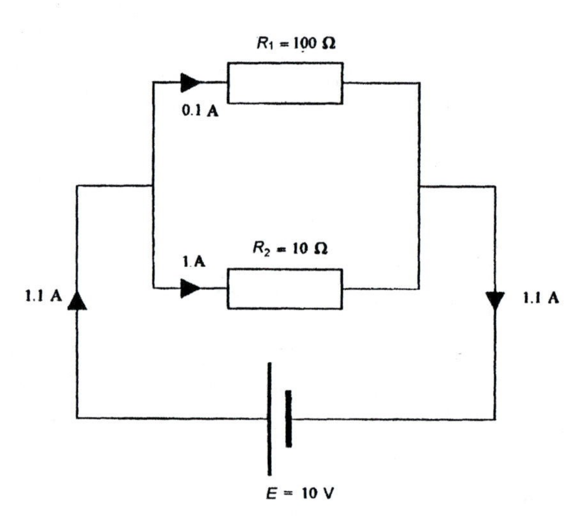

Figure 6.10
Current divides by a proportion of 10:1

In some instances it may not be easy to determine the total resistance of a parallel circuit by using proportion. In such cases we have to calculate the total resistance using the formula:

$$\frac{1}{R_t} = \frac{1}{R_1} + \frac{1}{R_2} + \frac{1}{R_3}$$ (assuming there are three resistors connected in parallel)

Note that this equation gives us the *inverse* of the total resistance, that is, $1/R_t$. To find R_t we must take the reciprocal.

Example

For the circuit in Figure 6.11, find the total resistance.

Solution

There are two ways of approaching this. We could find the common denominator of 2k, 1k, and 100 (which is 2k), and work the sum out long hand. Or, if we are using a calculator, we can simply calculate the values of $1/R_1$, $1/R_2$, and $1/R_3$ and add them together. Using the second method:

$1/R_t = 1/100 + 1/1000 + 1/2000$
$= 0.01 + 0.001 + 0.0005$
$= 0.0115$

If $1/R_t = 0.0115$, then R_t must equal the reciprocal of this so

$R_t = 87\ \Omega$

Figure 6.11

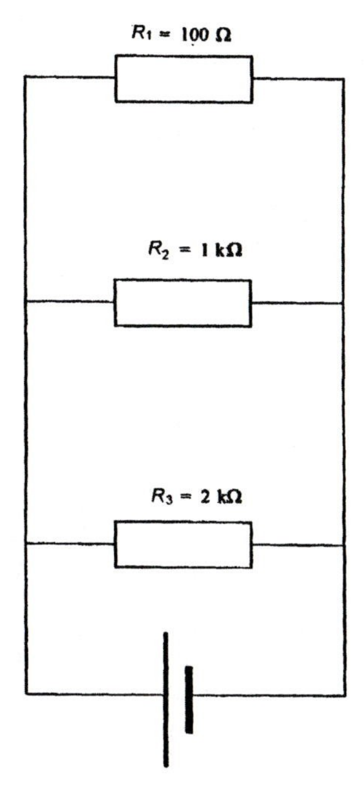

The above equation works for any number of resistors connected in parallel; however, where there are only two resistors connected in parallel an alternative equation *may* be used:

$$R_t = \frac{R_1 \times R_2}{R_1 + R_2}$$

This method of often referred to as *product over sum.*

Series/parallel networks

In a circuit, resistors (or components containing resistance) are connected in both series and parallel. There will be occasions where it is necessary to analyse these circuits to find either the total resistance, or the voltage, current, or power in part of the circuit.

The key to calculating the total resistance of a circuit is to identify the sections that have resistors in series, resolve these first, and then work out the parallel circuits.

Example

Calculate the total resistance of the circuit in Figure 6.12(a).

Solution

First resolve the series circuit of R_1 and R_2, where $R_1 + R_2 = 200\ \Omega$. The circuit can now be redrawn as shown in Figure 6.12(b), where R_1 and R_2 are replaced by an equivalent resistance labelled R_{12}.

Second resolve the parallel circuit of R_{12} and R_3, where, because both resistances are the same value, we know that the total will be half. Thus in Figure 6.12(c) we have an equivalent circuit with R_4 in series with the 100 Ω resistance R_{123}.

Finally we find R_t by resolving the series circuit. Thus $R_{123} + R_4 = 600\ \Omega$.

Potential divider

We know that when two resistors are connected across a voltage, a p.d. is developed across each resistor. This arrangement is known as a *potential divider* and is used to provide different voltages from a single source.

A potential divider is shown in Figure 6.13 where 10 V dc is applied across two 500 Ω resistors. When measured with respect to the negative line (C), the potential at (A) will be 10 V. The potential at (B) will be the p.d. across R_2 which, because the resistors are of equal value, will be half the input voltage; i.e. 5 V.

Where the resistors are of different values the potential at the centre will depend upon the ratio of $R_1{:}R_2$. Where R_2 is smaller in value than

R_1 the potential will be less than half the input. Where R_2 has a value greater than R_1 the potential will be greater than half the input.

Figure 6.12

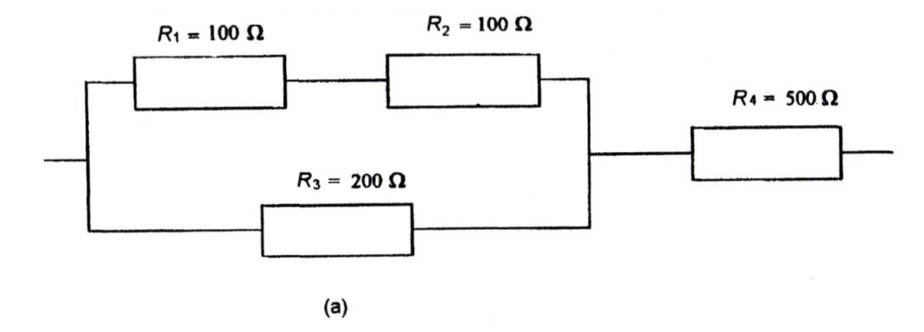

(a)

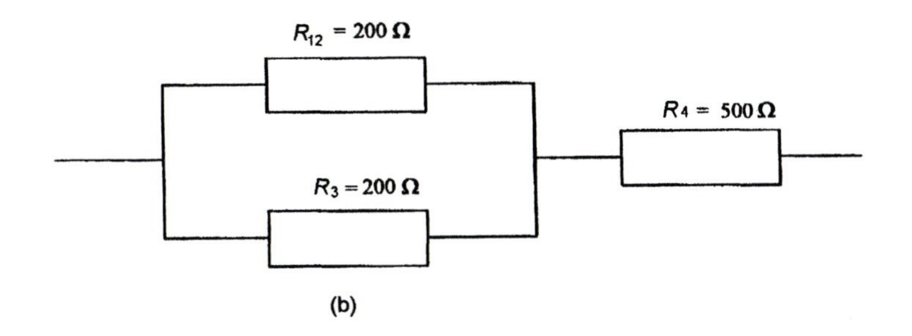

(b)

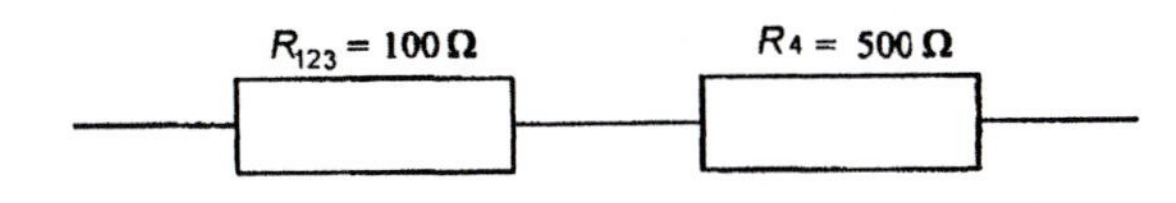

(c)

Figure 6.13
A potential divider

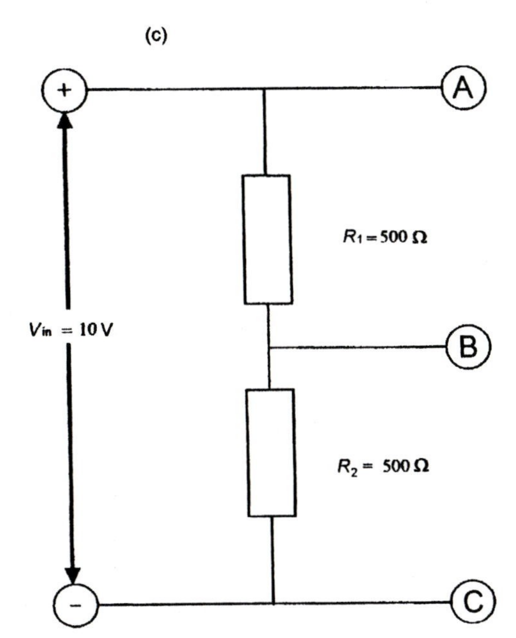

To *calculate* the p.d. across one of the resistors we would apply Ohm's law. But this leads us to the following equation:

To find VR_1, we would use $I \times R_1$.
But $I = V_{in}/R_t$.

$$\text{Thus, } VR_1 = \frac{V_{in}}{R_1 + R_2} \times R_1$$

Transposing the equation we have

$$VR_1 = V_{in} \times \frac{R_1}{R_1 + R_2} \quad \text{and} \quad VR_2 = V_{in} \times \frac{R_2}{R_1 + R_2}$$

Example

In a potential divider like that shown in Figure 6.13, R_1 has a value of 10 kΩ, R_2 has a value of 1k5 Ω, and the input voltage is 150 V. Find the potential at point B.

Solution

$$VR_2 = \frac{150 \times 1.5 \times 10^3}{10 \times 10^3 + 1.5 \times 10^3} = \frac{150 \times 1.5 \times 10^3}{11.5 \times 10^3} = \frac{225 \times 10^3}{11.5 \times 10^3}$$

Thus $VR_2 = 19.6$ V.

We can now see that by changing the values of the two resistors in a potential divider we can derive any value of output voltage between V_{in} and 0 V. This principle has many applications in electronics.

Variable resistors

These are called *potentiometers* because they use the potential divider principle.

A carbon track is used to create a specific value of resistance. Terminals are attached to each end of the track, and a sliding contact is made to travel along it. The principle is shown in Figure 6.14(a), where for simplicity the three terminals labelled 'A', 'B', and 'C' relate directly to Figure 6.13. If the track is made to have a resistance of, say, 1 kΩ, the resistance between terminals 'B' and 'C' will have a value of anything between 1 kΩ and 0 Ω, depending on the position of the slider.

A straight track means that the control will be adjusted by sliding the control knob. Potentiometers of this construction are rather large, and are used mainly for customer controls in domestic audio and TV equipment.

An adaptation of this control is the preset slider. In this device the sliding contact shown in Figure 6.14(a) is attached to a threaded rod, and the contact moves along slowly when the rod is rotated. The thread is geared so that up to 30 turns are required to allow one travel of the contact, allowing very fine adjustments to be performed. These devices were used extensively in television tuning circuits before being replaced by microprocessor arrangements, and they are now found mainly in industrial equipment.

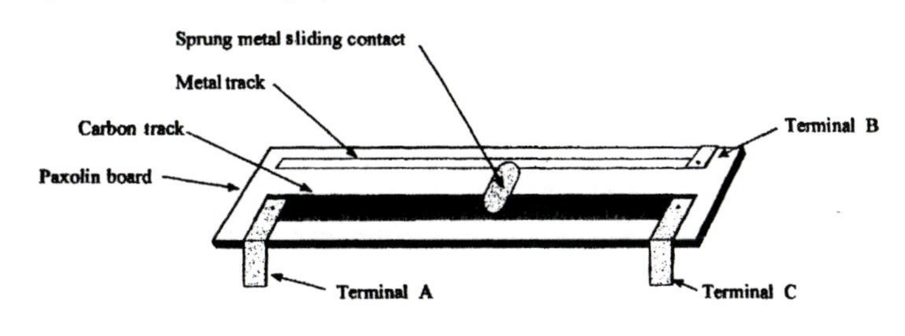

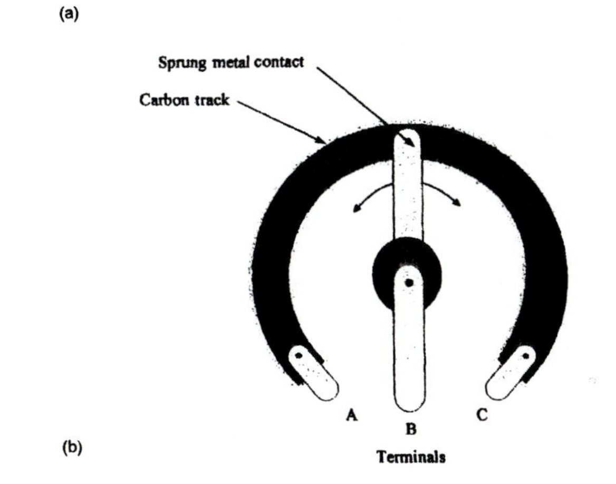

Figure 6.14 (a) Sliding construction carbon potentiometer. (b) Rotary construction carbon potentiometer

By folding the track as shown in Figure 6.14(b) the control can be made into the more common rotary type. For user control applications in electronic equipment a shaft may be fixed to the wiper to enable customer accessibility. Some examples are TV user controls, VCR tracking control, photocopier darkness, etc.

Skeleton potentiometers are designed to be used inside equipment and are only accessible to the manufacturer or service engineer. These variable resistors are used to compensate for manufacturing tolerances introduced by individual component tolerances, and when adjusted correctly will ensure that each piece of equipment that comes off a production line performs to the same specification. There are occasions where a service engineer has to adjust these controls in the course of repair and maintenance, but they can be abused by some who turn them at random to 'see what happens'! This is not acceptable practice.

Another common application of the potentiometer is in joysticks. Here two controls, usually of the rotary type, are mounted at right angles to each other. The stick is mechanically connected to these controls in such a way that forward and back movement causes one control to rotate, whilst side to side movement causes the other control to rotate. The equipment connected to the joystick, e.g. a computer, can ascertain the position of the stick by reading the value of the two variable resistors. The principle is shown in Figure 6.15.

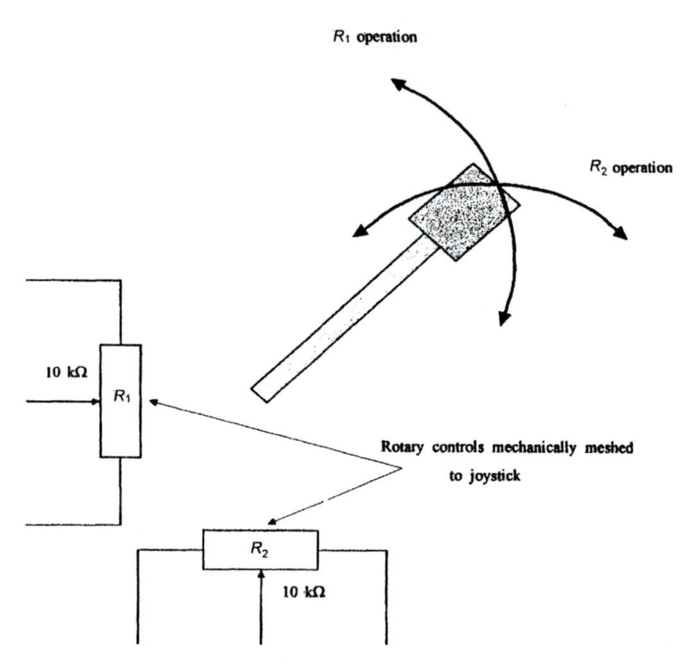

Figure 6.15
Potentiometer configuration for a joystick

The BS 3939 symbols for potentiometers are shown in Figure 6.16.

There are two track designs for carbon potentiometers, linear and logarithmic.

In a linear control, the carbon track is made so that each equal movement of the slider gives an equal change in resistance. It is this type of potentiometer that is used most commonly in equipment.

In a logarithmic control the resistance of the track changes in a non-linear fashion for each equal movement of the slider. At the start of rotation the resistance change per movement of slider is small, increasing as the slider moves along the track. The rate of change of resistance follows a logarithmic rule, and because the human ear follows a logarithmic rule, these are commonly used for volume controls in audio equipment, the characteristic of the control matching that of the ear.

Figure 6.16

British Standard symbols denoting various types of potentiometer

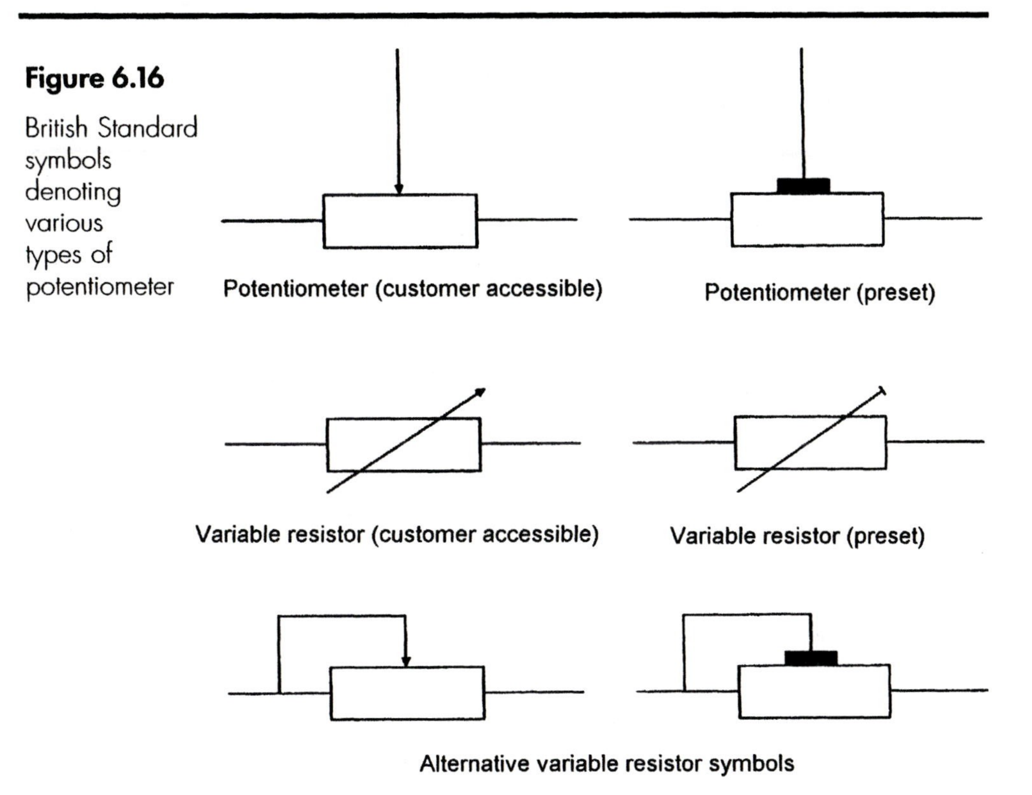

As with carbon resistors, potentiometers dissipate an amount of power. For most applications this is low, but where a power level of a few watts is expected to be dissipated in a potentiometer, a wirewound construction is used. In this case the 'track' is made up from a spiral of resistive wire wrapped around a ceramic former. A sprung metal contact acts as a wiper.

Thermistors

This type of resistor is designed specifically to take advantage of the effects of temperature coefficient. Unlike most other resistors which are carbon based, thermistors are semiconductor based, and are designed to change their resistance with temperature.

There are two types: the *p.t.c.* whose resistance increases with a rise in temperature, and the *n.t.c.* whose resistance decreases with a rise in temperature.

Some typical applications of thermistors are temperature sensors in heating systems, heat sensors in fire detection equipment, and thermal switches in colour TV degaussing (tube demagnetising) circuits.

Thermistors are also commonly used for thermal compensation in electronic circuits, where the unwanted effects of temperature coefficient on individual components are corrected by including a thermistor in the circuit design. For example, a component which suffers a fall in resistance as it heats up during operation could be compensated for by inserting a p.t.c. When the circuit is cold the component resistance will be high and the p.t.c. low. As the circuit warms up the component resistance will fall, but the p.t.c. will rise to compensate.

The BS 3939 symbols for thermistors are shown in Figure 6.17.

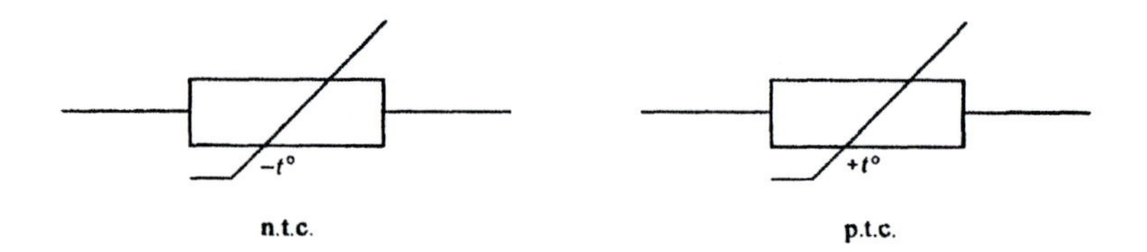

Figure 6.17 British Standard symbols for thermistors

Voltage dependent resistors

Commonly referred to as a V.D.R.s, these are another form of semiconductor resistor which exhibit a change in resistance when the voltage applied across the device changes. The BS 3939 symbols are shown in Figure 6.18.

Figure 6.18 British Standard symbols for V.D.R.s

V.D.R.s have a number of applications in electronic circuits. One example is noise spike suppression.

In Figure 6.19(a), the V.D.R. has a high resistance at the operating voltage, so only a low current flows through the V.D.R.

In Figure 6.19(b), when a noise spike of 1 kV appears on the line, the resistance of the V.D.R. falls, the V.D.R. current rises dramatically, and the noise spike is damped. Thus the V.D.R. is acting as a protection for the load circuit which may easily be damaged by such a large noise spike.

Figure 6.19
V.D.R. as a noise suppressor

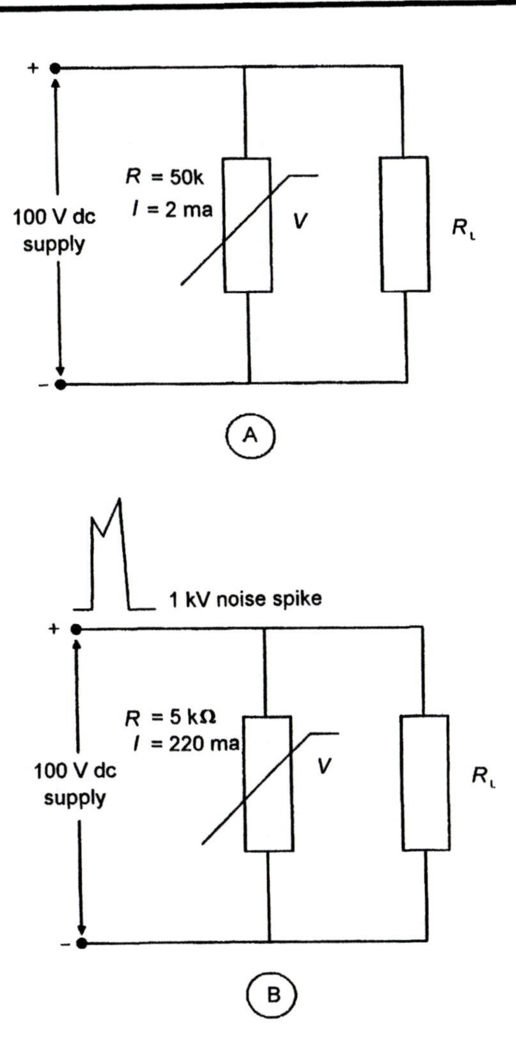

7

Chemical cells

When suitable electrodes are immersed in an electrolyte an e.m.f. is produced between them. When the electrodes are connected to an external electrical circuit a current flows. At the same time chemical action takes place at the electrodes. The device acts as a source of e.m.f. and is known as an *electric cell.*

A single cell will only provide a very small e.m.f., and has a limited current capacity. For this reason a number of cells may be connected together in either series or parallel. When connected in series, the e.m.f. of each cell is added to its neighbour to give a total e.m.f. In parallel connection the e.m.f. is unchanged; however, the total current capacity will be equal to the addition of the capacity of each cell. Series and parallel connections are shown in Figure 7.1.

The devices sold in the high street, commonly known as batteries, are in truth cells, having an e.m.f. typically in the order of 1.2 V–1.5 V. A battery is a cluster of cells connected (usually) in series to give a greater output e.m.f. For example, a 9 V battery would be constructed from six cells each having an e.m.f. of 1.5 V. The physical size of a 9 V battery gives some indication of the current capacity. A larger battery will be made up of six larger cells thus offering the option of a low current drain over a prolonged period, or a high current drain over a much shorter period.

There are two types of cell: *primary* and *secondary.*

Primary cells cannot be recharged to any useful extent once the chemicals are exhausted and are considered disposable. Most manufacturers warn that any attempts to recharge a primary cell could be dangerous as they are liable to explode due to a build-up of gas pressure inside them. There are some charging circuits that claim to be able to recharge primary cells, and although they do have some degree of success the duration of a primary cell cannot match that of a secondary cell. *A primary cell must never be placed in a charging unit designed only for rechargeable cells.*

Figure 7.1
Cells connected to form batteries

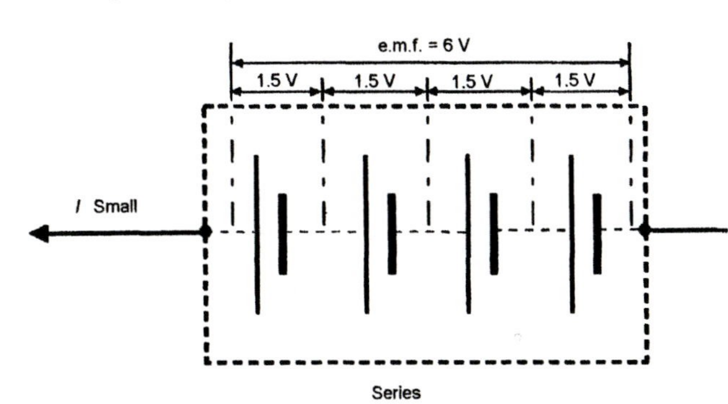

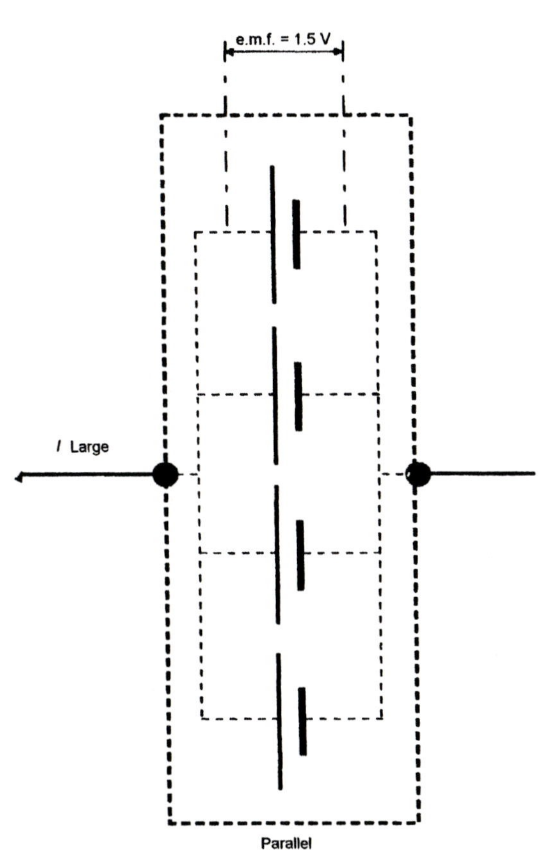

The term *secondary cell* is given to any cell that is rechargeable. For many years the most common applications were in motor vehicles (including milk floats), telephone exchanges, emergency lighting systems, and some portable equipment such as mobile radio. However, with the increase in both industrial and domestic portable items, many of which have a high current consumption, the demand for cheap rechargeable cells and batteries has risen sharply since the 1970s. In the majority of cases a secondary cell will perform equally as well as its primary counterpart, and although the initial purchase price is somewhat higher, this is quickly recouped because it will recharge hundreds of times if correctly used.

One drawback with secondary cells is that they self-discharge much more rapidly than primary cells, which does not make them an ideal choice for devices such as clocks, watches, smoke alarms, or any other device where a constant trickle current is necessary. Primary cells are more suited to such applications.

Primary cell construction

A simple primary cell is illustrated in Figure 7.2.

Figure 7.2
Principle of a simple voltaic cell

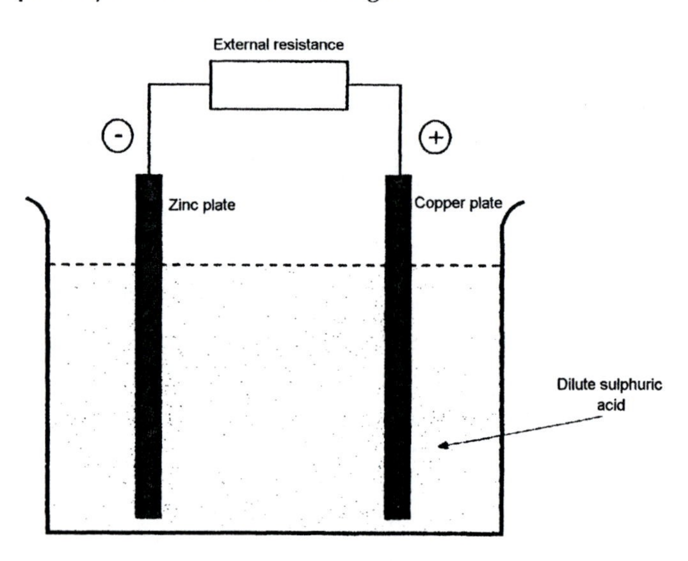

In this device, zinc is dissolved at the negative plate to form zinc sulphate, and hydrogen is formed at the copper positive plate. This cell as an e.m.f. of about 0.7 V.

When a current flows out of the cell, hydrogen bubbles form at the positive plate. Meanwhile the zinc plate is attacked by the acid, forming zinc sulphate. The cell has a limited life because of these effects.

In practice the amount of energy which can be supplied by the cell is quite small because the hydrogen bubbles produced at the copper electrode act in two ways to reduce the current flow. First of all they mask the electrode and prevent ions reaching it in large numbers. Secondly, the copper, hydrogen, and acid collectively act as a cell and set up an e.m.f. in opposition to the main cell. This combined effect, known as *polarisation,* reduces the effectiveness of the cell to practically zero in a very short time. For the cell to be of any practical use some means of overcoming the problem of hydrogen build-up must be used.

Although commercial cells employ the principle of the simple voltaic cell shown in Figure 7.2, a depolarising agent is added which removes the hydrogen bubbles by oxidising it to water. Also, different chemicals are used which give a higher terminal e.m.f.

For many years mercury was included in the production of cells; however, this substance is highly toxic and is considered too dangerous to be thrown into landfill sites. With the number of cells and batteries that were being disposed of each year, it became evident that the use of mercury in dry cells should be curtailed.

Dry cells

The Leclanché dry cell was for many years sold in the high street as a disposable cell or battery. In its earlier form it comprised of a carbon rod which formed the positive pole, and a zinc case which was the negative pole. To improve performance, mercury was included in the manufacture of these cells; however, as the need to remove mercury became more evident, new improved cells were developed, and the Leclanché evolved into the zinc chloride cell. With a similar performance to the zinc carbon Leclanché, this cell offers a reasonable shelf life (2 years) and capacity at an affordable price.

Where higher capacity is required, the alkaline manganese cell has become the dominant disposable cell in the high street because it offers good leakage resistance, has a long (5 year) shelf life, and functions well under both high and low current drain conditions.

In all of these cells the electrolyte is a moist jelly with a depolariser surrounding the centre rod, removing the hydrogen which forms in this area. The basic construction is illustrated in Figure 7.3. The terminal e.m.f. is typically 1.5 V; however, this may fall after a period on load, but the cell will return to normal if it is 'rested', giving the depolariser time to act. As the cell ages, the length of time that it can perform before its output falls becomes shorter as the depolarising agent deteriorates.

Figure 7.3
Construction of a Leclanché cell

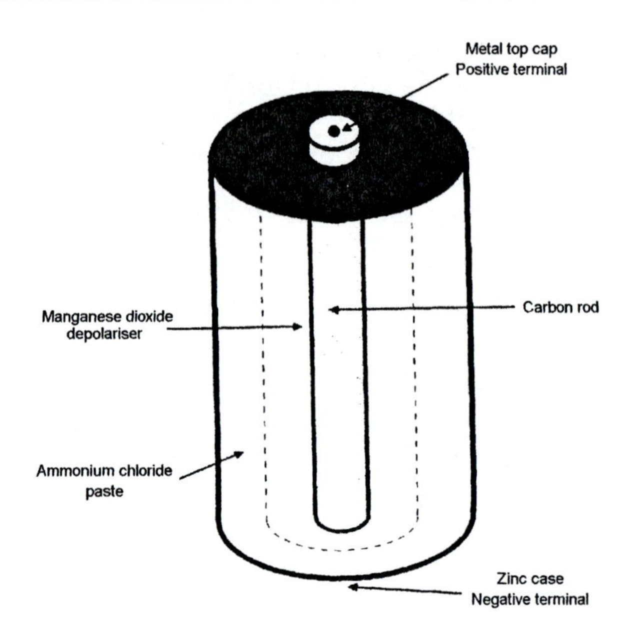

The mercury cell

This type of cell was developed to meet the requirements of miniaturisation, e.g. guided missiles, medical electronics, hearing aids, etc. Its output voltage varies by less than 1% over a period of months, and remains stable over a wide range of temperature and pressure. Furthermore, due to its superior depolarising action, it does not require rest periods, it has a long shelf life, and can be made to be completely leak proof, as well as resistant to humidity and corrosion.

The cell has a zinc anode, a mercury oxide cathode, and a manganese dioxide depolariser. It has a double steel case. The terminal voltage is in the order of 1.35 V.

Various forms of construction are used, depending on the application and required capacity. A button construction is shown in Figure 7.4.

Despite their high performance characteristics, these cells are being replaced with more environmentally friendly devices.

The lithium cell

The most outstanding factor in the characteristics of lithium cells is their long life, which can be up to ten years. Depending on the chemical composition used, a lithium cell can offer a low current performance over a prolonged time period, or a higher current output over a shorter time scale, although the ten year shelf life is still retained.

Figure 7.4
Mercury cell construction

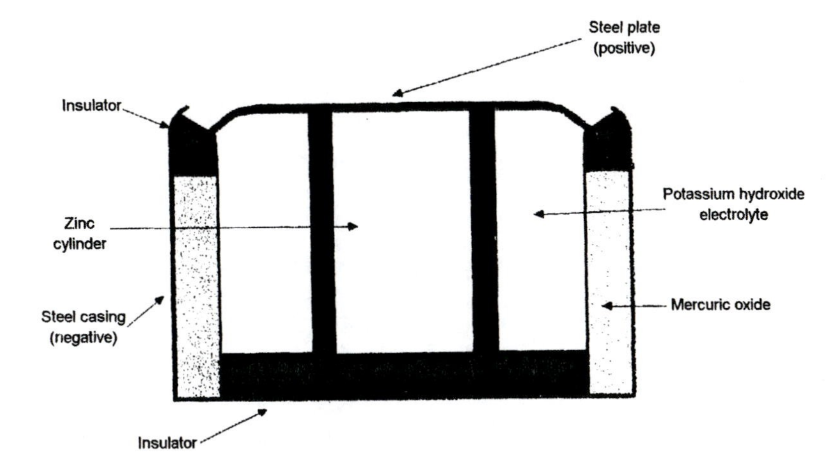

The long life makes these cells ideal for use in memory chip power supply back-up applications. For example, in timers used in VCRs, ovens, etc., and computer clock back-up supplies to maintain the clock function when the mains power is removed.

The cell voltage is dependent upon the chemical composition used; for example, a lithium iron disulphide cell has a terminal voltage of 1.5 V, whereas a lithium thionyl chloride cell has a terminal voltage of 3.7 V. A button cell version is available with a terminal voltage of 3 V.

The Lithium cell is a primary cell; however, there is a vanadium lithium version which is a rechargeable (secondary) cell.

The nickel cadmium secondary cell

The nicad, as it is commonly called, best meets the demand for a relatively cheap, high capacity cell or battery to operate the sort of portable equipment which is in common use. It can supply a far higher constant current than its equivalent size primary cells and, if properly cared for, it can be recharged many hundreds of times. The construction of a typical cell is shown in Figure 7.5.

The positive electrode is a nickel hydroxide paste, and the negative a cadmium paste. The electrodes are formed by binding the paste into a steel mesh, and are insulated by a separating layer. The electrolyte is formed from a potassium hydroxide gel. When fully charged the anode is nickel and the cathode is cadmium, the materials changing to nickel oxide and cadmium hydroxide when discharging. The nicad cell has an e.m.f. of 1.3 V.

Figure 7.5
Cutaway view of a nickel cadmium cell

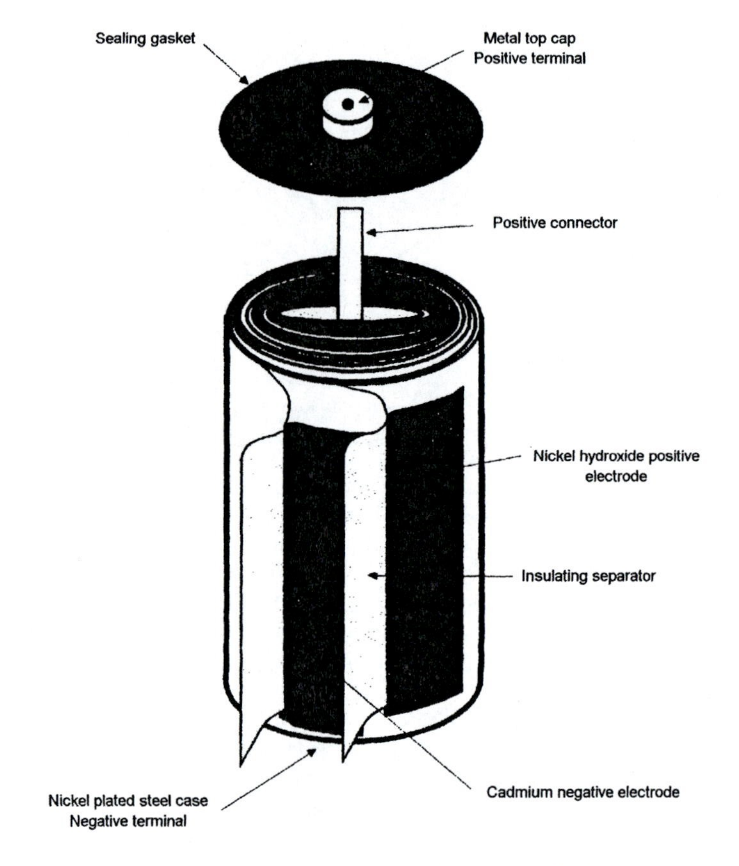

There are two important factors which govern the life of a nicad cell: the rate at which it is charged, and the condition in which it is stored when not in use.

The charging rate is controlled by using a constant current source which is designed to limit the initial charge rate, thus preventing overheating.

Storage is the responsibility of the user, but they should be made aware of the fact that storing nicads in a semi-charged state will shorten the life of the cell. Manufacturers generally recommend that nicads should be taken through a rapid charge and discharge cycle every few months when not in use. For this reason many domestic camcorders employing nicads incorporate a discharge circuit; alternatively discharge units can be purchased separately.

A nicad performs well as a back-up battery if it can be maintained on a constant charge. Examples of such applications are found in security alarm bell modules, and microprocessor memories such as those found in VCR timers and television tuning circuits.

Lead-acid cells

These are made up of a lead peroxide positive plate, a lead negative plate, and a sulphuric acid/water electrolyte. During discharge both plates turn to lead sulphate due to absorption of the acid; however, when an e.m.f. is applied across the terminals the process reverses and the cell is once more fully charged. They have a nominal e.m.f. 2.3 V.

The condition of a lead-acid cell can be ascertained by measuring the density of the water in the electrolyte because, as the cell discharges and the acid is absorbed, the density of the water reduces. The density is referred to as the specific gravity (SG), and is measured using an instrument called a hydrometer. The hydrometer is floated in the electrolyte (which is extracted from the cell using a special sealed syringe) and a reading taken from the scale on the side. Typical SG readings are 1.28 for a fully charged cell, and 1.16 for a fully discharged cell.

For most applications a single cell is not sufficient, and so lead-acid batteries are more common, offering e.m.f.s typically in the order of 6 V and 12 V. The water in the electrolyte can evaporate and must therefore be topped up. Caution must be taken during this operation as the acid is highly corrosive, and the fumes given off are very toxic. For this reason lead-acid batteries must be operated in suitably ventilated conditions.

A sealed version of the lead-acid battery is available. In this case the electrolyte is suspended in a glass fibre material which prevents leakage. The gas discharge is cut by incorporating a method of recombining up to 99% of the gas generated. A self-resealing valve in each cell ensures that any build-up of gas pressure will be released automatically.

The advantage of the lead-acid cell over its nicad counterpart is its much lower manufacturing cost. A lead-acid battery can provide current in the order of hundreds of amps without risk of damage, and can be quickly recharged again. The charging requirements are less complex because lead-acid cells require a constant voltage which is more easily provided.

Because of their high current capability coupled with their relatively low cost, lead-acid batteries are used extensively in industry for such applications as emergency lighting, security and fire alarm back-up supplies, portable power tools, portable medical equipment, etc.

And, of course, motor vehicles rely on lead-acid batteries for their source of electrical power.

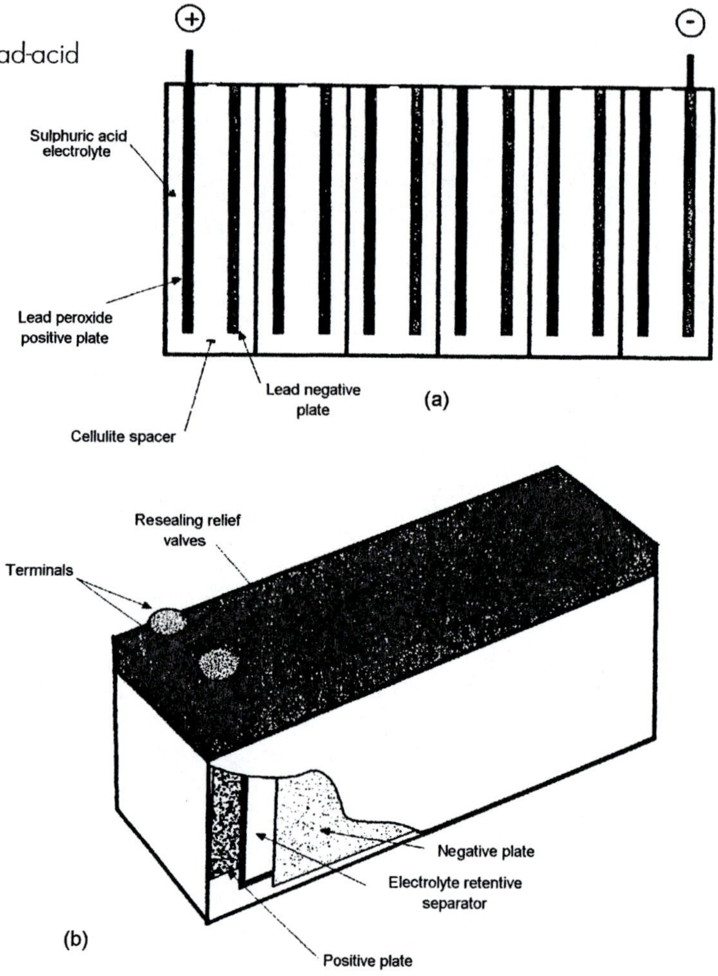

Figure 7.6
(a) Six 2.3 V lead-acid cells forming a 13.7 V (commonly called 12 V) battery. The plates would all be internally connected in series. (b) Cutaway view of a sealed lead-acid battery

Internal resistance

When electrolysis takes place the main body of the electrolyte between the electrodes acts as a conductor. The e.m.f. of the cell is generated within the vicinity of the electrodes. Energy is required to move the ions through the electrolyte, and thus the cell can be said to have a resistance. This is known as the internal resistance and must be distinguished from any external resistance in the circuitry.

An example is given in Figure 7.7 where a cell with an e.m.f. of e volts has a reduced terminal voltage because of the internal resistance R_{in}. Thus, from Figure 7.7 we see that the actual terminal e.m.f. will be:

$$E = e - V_{in}$$

However, V_{in} is not constant because it is dependent upon the current being taken by the external load. Furthermore, the value of R_{in} varies because it is dependent on the condition of the depolariser, which alters during use. In practice the e.m.f. quoted for a cell is that which is measured at the terminals of a fully charged cell.

Figure 7.7
Effect of battery internal resistance terminal voltage

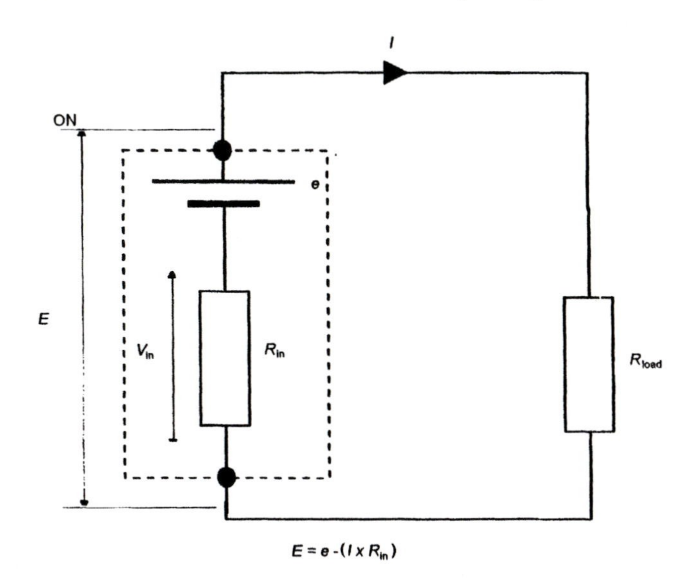

Example
A cell has a terminal e.m.f. of 1.24 V when supplying a load of 3 A. When the load is disconnected the cell e.m.f. rises to 1.28 V. Calculate the internal resistance of the cell.

Solution
Internal p.d. at 3 A = 1.28 V − 1.24 V = 0.4 V.

$$\therefore R_{in} = \frac{0.04\,\text{V}}{3\,\text{A}} = 0.0133\ \Omega$$

Cell capacity

This is stated in amperes per hour (Ah), and is the charge moved by a current of 1 A in 1 hour. For example, if a battery with a rating of 60 Ah is discharged at a constant rate of 6 A, the operating time of the battery would be found from the equation $Ah = I.t$.

$\therefore$ transposing for time; $t = Ah \div I$
$= 60 \div 6 = 10$ hours

A fully charged 6 Ah battery in good operating condition could in theory provide a current of 6 A for 1 hour, or 1 A for 6 hours. However, it does not necessarily follow that the battery could provide a current of 1 A without overheating and becoming permanently damaged, this would depend on the type and size of the battery.

In practice, when calculating the capacity of a battery required for a particular application, it is common to assume that the battery will be 80% efficient, so a battery of a slightly higher capacity than calculated is selected.

8

Capacitance

Capacitance is the property of storing an electrical charge. One component having this property is called a capacitor.

This device is constructed from two plates that are separated by an insulator, as illustrated in Figure 8.1. The insulating material, called the dielectric, can be anything such as a thin piece of paper, mica, or ceramic, or simply a small air gap. The use of different dielectric materials will be considered later.

Figure 8.1
Basic capacitor construction

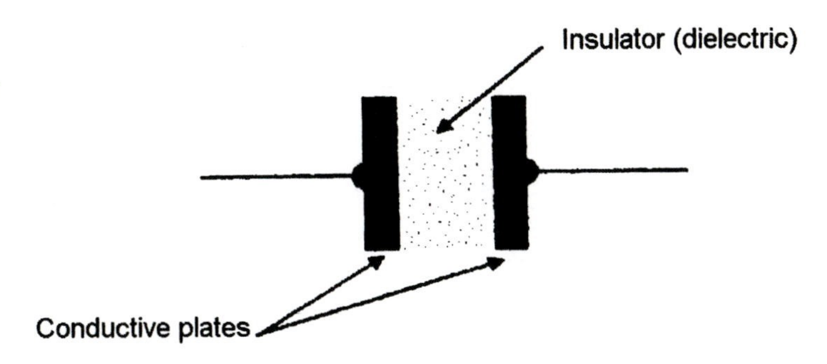

The operation of a capacitor is shown in Figure 8.2(a) and (b).

In Figure 8.2(a), when switch S_1 is closed the positive electrode of the battery attracts electrons from plate A of the capacitor, leaving it with a deficiency of electrons and thus in a state of positive charge. Plate A will now attempt to satisfy this charge by drawing electrons from the negative pole of the battery, but because they cannot penetrate the dielectric they collect on plate B making it negatively charged. This charging current will continue to flow until the charge on the capacitor equals the applied e.m.f.

When S_1 is opened the charge on the capacitor will in theory remain unchanged because the electrons will not have any path through which they can move. In practice a capacitor has some leakage path through the dielectric causing it to lose its charge slowly over a period of time.

In Figure 8.2(b) we see the effect on the circuit when the switch is closed, and the charged capacitor has been placed across a load resistor. The capacitor now acts as an e.m.f. where electrons on plate B are attracted to the positive charge on plate A. Current will continue to flow until the charge on both plates is equal (zero volts across the capacitor). As with any circuit, the size of the current will depend on the value of R_l, and the amount of charge on the capacitor.

Figure 8.2

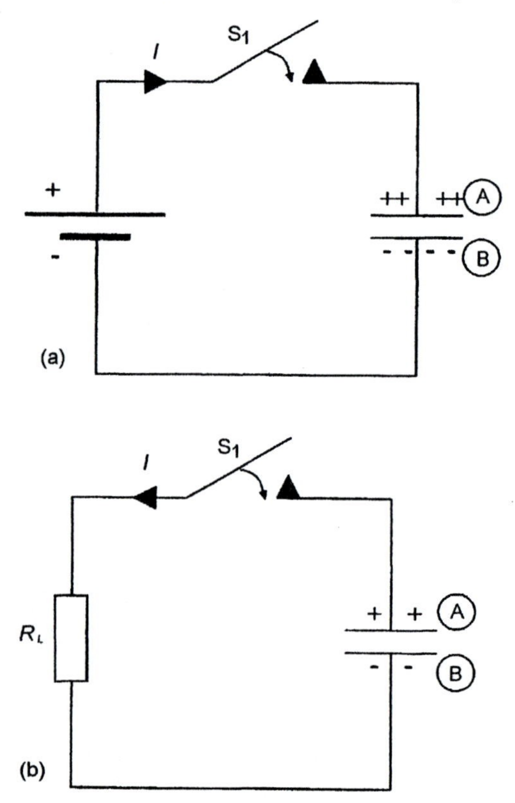

Value of a capacitor

A capacitor's value is determined by the amount of charge that it can store. This is governed by three factors: *the area of the plates, the distance between the plates* (in other words the thickness of the dielectric), and *the type of dielectric*.

Area of the plates. It is obvious that the greater the plate area, the more space there is to store electrons. So the greater the area, the higher the value.

Distance between the plates. If we consider Figure 8.2(a) again, we can see that electrons will only be attracted from the battery if there is a strong enough electric field from the positive plate of the capacitor, and the thicker the dielectric, the weaker the field will be. So the shorter the distance between the plates, the higher the value.

Type of dielectric. The electric field formed by the charge on the capacitor must pass through the dielectric (Figure 8.3). But the total amount of flux will depend on the opposition of the dielectric to the flux. This opposition is known as elastance. This quantity is similar in behaviour to the resistance in an electrical circuit, and clearly if the dielectric elastance is high, then there will be less attraction of electrons from the battery by the positive plate in the capacitor. The properties that determine the elastance of a dielectric are quantified as *relative permittivity*.

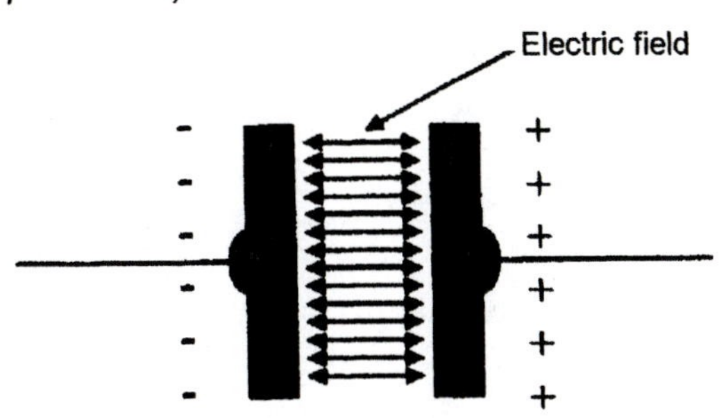

Figure 8.3
Electric field acting between the plates of a charged capacitor

These factors governing the value of a capacitor are expressed in the formula:

$$C = E_o.E_r \times \frac{a}{d}$$

where E_o = the relative permittivity of free space, E_r = the relative permittivity of the dielectric, a = the area of the plates and d = the distance between the plates.

It is true to say that if we ignore the effects of the dielectric permittivity, then the value of a capacitor is directly proportional to the area of the plates, and the distance between them. Thus $C \propto a/d$.

Charge in a capacitor

The size of the charge in a capacitor is proportional to the applied e.m.f., and the value of the capacitor. You will recall that electrical charge is measured in coulombs (Q), therefore the charge in a capacitor can be found from:

$$Q = V.C$$

where Q = electrical charge, V = applied e.m.f. and C = capacitance value.

This equation assumes that the capacitor has been permitted to charge fully. However, it takes a certain amount of time for the electrons to move into the capacitor, the time period being dependent on the amount of current flowing. Therefore if the charge in a capacitor is a function of both time and charging current, then the charge can be found from:

$$Q = I.t$$

where I = charging current and t = time in which charging takes place.

The relationship between these two equations will be examined in more depth later on when considering time constant.

Electrical charge can be related to energy, and the amount of energy stored in a capacitor can be found from:

$$\omega = 0.5 \times C \times V^2$$

where ω (omega) = energy, C = capacitance value and V = applied e.m.f.

The unit of capacitance is the *farad (F)*. If a capacitor stores a charge of 1 coulomb when an e.m.f. of 1 volt is applied across it, the value of the capacitor is said to be 1 farad.

The coulomb is a very large unit (a value of 6.3×10^{18} electrons in 1 second), and because the farad is related to the coulomb this too is a very large unit. In practice capacitor values in the order of farads are not required, and fractions of one farad are used; i.e. microfarads (μF; 10^{-6} farads), nanofarads (nF; 10^{-9} farads), and picofarads (pF; 10^{-12} farads).

Dielectric strength

As the e.m.f. across the plates is increased, the atoms which make up the dielectric are placed under stress. Electrons on the negative plate will attempt to break through the dielectric to reach the positive plate, and when the e.m.f. becomes too high, breakdown occurs and the capacitor conducts. The dielectric is thus destroyed and the capacitor is said to be a short circuit if the resistance of the dielectric is 0 Ω, or leaky if the dielectric still has some value of resistance.

Because of the limitations of the dielectric, manufacturers specify voltage ratings for capacitors, and this rating must not be exceeded.

Dielectric loss

The stress set up in a dielectric requires a certain amount of energy to be expended, and this energy appears as heat. Clearly the higher the stress, the greater the losses in the capacitor, which leads us to two important considerations when selecting a capacitor for a particular application: first, if the capacitor is running warm it is more liable to failure due to dielectric breakdown; second, unnecessary heat loss within a circuit equates to inefficiency and undue power consumption. For these two reasons it is important that the voltage rating of a capacitor is somewhat greater than the maximum voltage that will be applied by the circuit. A good rule of thumb is to rate the capacitor at least one third higher than the maximum applied voltage.

Another important point to note is that the dielectric loss increases as the frequency of the charge/discharge rate rises. In other words, if ac signals are applied to a capacitor, the losses will rise as the frequency of the ac increases.

Capacitive reactance

A capacitor has a high resistance when dc is applied. Look again at Figure 8.2(a). Although an initial current flows when the switch is closed, once the capacitor is fully charged current flow stops, and the resistance in the circuit becomes the resistance of the dielectric, which is extremely high.

However when, as in Figure 8.4, an ac signal is applied, the capacitor will charge to one polarity during the positive half cycles, and will then discharge and charge up in the opposite polarity during the negative half cycles. This means that there is always a current flowing around the circuit, and thus the capacitor can be considered to have a different value of resistance to ac than to dc. If the applied frequency increases, the rate of charge/discharge increases and the circuit current rises accordingly. By Ohm's law definition, if the circuit current increases when there has been no increase in applied e.m.f., then the circuit resistance must have reduced. Thus, as the applied voltage frequency across a capacitor increases, its apparent resistance falls.

A capacitor's opposition to current flow is not constant but varies with the frequency applied. The opposition to current flow is termed *reactance* (X_C), and the unit of the ohm is used. In a capacitor, the reactance is high at low frequencies, and low at high frequencies. This is illustrated by the graph in Figure 8.5.

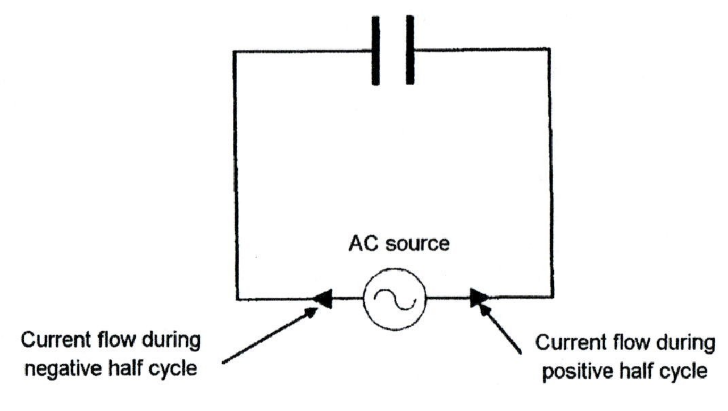

Figure 8.4 Capacitor behaviour in an ac circuit

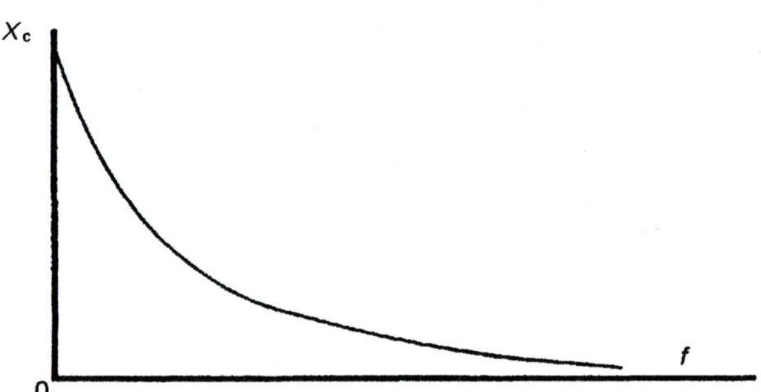

Figure 8.5 Change in capacitive reactance with frequency

It is not only the applied frequency that affects the reactance of a capacitor. The value of the capacitor also has a bearing because, for any given frequency, as the capacitive value increases, so the charge/discharge current increases, indicating a fall in reactance.

The reactance of a capacitor can be calculated from the equation:

$$X_C = \frac{1}{2\pi.f.C} \text{ ohms}$$

where f = applied frequency and C = capacitor value.

Capacitor construction

There are numerous types of insulator that may be used as a dielectric, and the choice of dielectric is governed by a number of considerations including the desired working voltage (dielectric strength), the physical size, range of operating temperatures, tolerance, and cost. In simple terms capacitors can be divided into the following categories; those that use a solid, dry dielectric, those that employ a 'wet' electrolyte, and those that use an air gap.

A common method of constructing dry dielectric capacitors is to take a thin sheet of either paper, polyester, polythene, or polycarbonate and coat both sides with silver. This gives the required two conductors with a separating insulator. To reduce the size to a practical dimension, the sheet is rolled like a Swiss roll, connecting terminals are attached, and the assembly is encapsulated in a plastic container sealed with epoxy resin, or is dipped into wax or plastic to provide both electrical insulation and keep out moisture. The process is shown in Figure 8.6, along with the standard BS 3939 symbol for a capacitor.

Figure 8.6
Capacitor construction and types of encapsulation

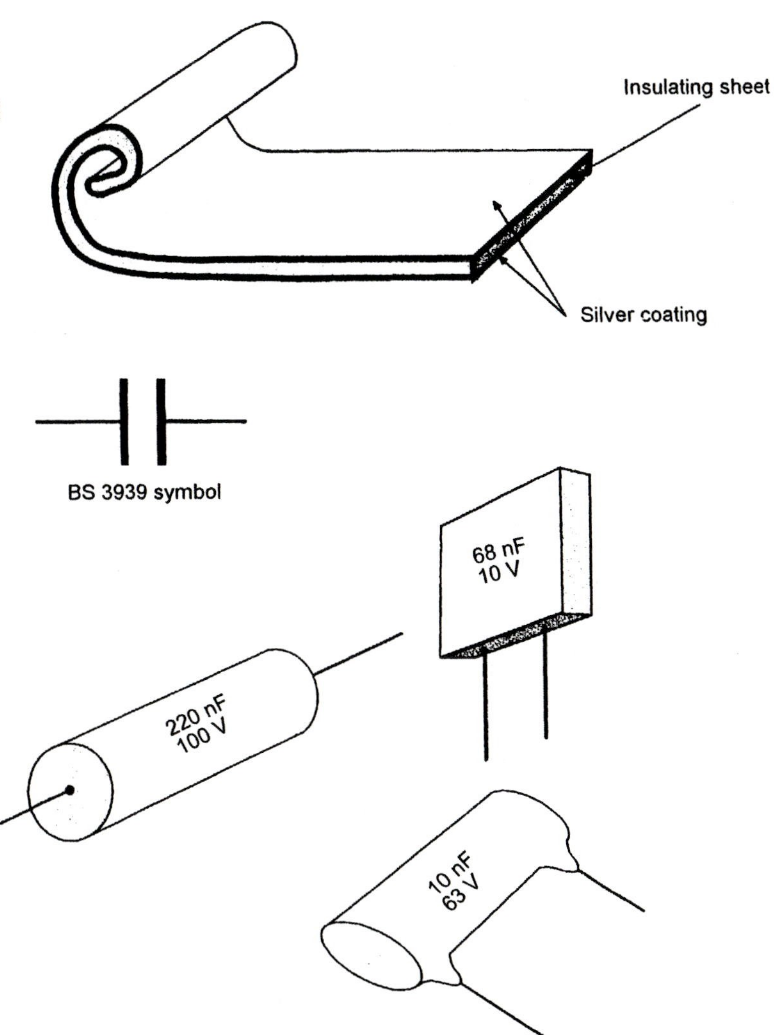

Capacitors come in a vast range of sizes, shapes, values, and working voltages. They are used in many types of electronic circuit including audio and radio frequency amplifiers, oscillators, and filters.

The rolled construction offers reasonably high values of capacitance and working voltages; however, once they exceed three or four microfarads in value their physical size is too large to fit into small equipment.

For very small values in the order of picofarads and a few nanofarads, disc construction is cheap and effective. Here a disc of mica, or ceramic is coated on both sides with silver or aluminium, contacts are attached, and the assembly is dipped in wax, plastic, or other insulating material. A typical example is shown in Figure 8.7.

Capacitors with low values are mainly used for high frequency applications such as radio tuning and filtering.

Figure 8.7
Disc ceramic capacitor

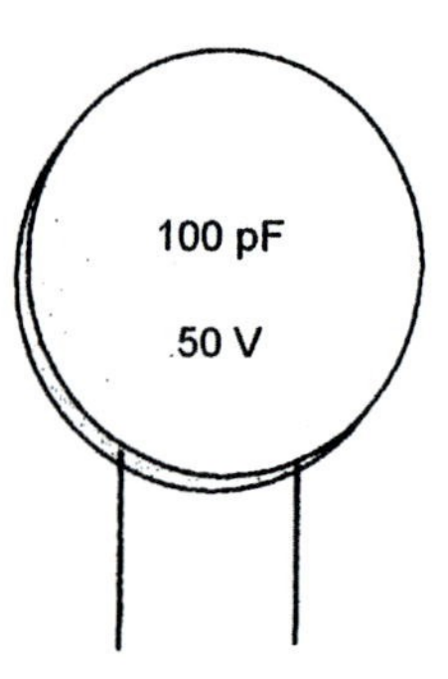

For capacitance values above one microfarad, wet *electrolytic* types are used. The construction is shown in Figure 8.8. Two sheets of aluminium foil are separated by a sheet of paper that has been impregnated by an electrolyte gel. One sheet of foil forms a positive plate whilst the gel in the paper forms a negative plate. The second sheet of foil is not actually an electrode at all, it simply serves to make a mechanical connection to the gel onto which to connect the terminal. The dielectric is formed during the manufacturing process by applying a voltage to the device which causes an aluminium oxide layer to build up on the positive plate. The assembly is placed inside a metal can, and a rubber bung is fixed into the end to seal the device.

Values of thousands of microfarads at hundreds of volts are possible at physical dimensions far smaller than would be possible using paper or similar dielectric materials. Electrolytics are used extensively for smoothing and filtering in power supplies, and as coupling devices in low frequency amplifiers.

One important point to note is that, unlike dry dielectric capacitors, electrolytics are polarity conscious, and if they are connected to a circuit the wrong way around they are liable to explode. Such an explosion can force the metal container to fly off with considerable force and can cause serious injury, especially if it catches the unwary engineer in the face. When fitting these devices, always ensure that you have connected it the correct way around. Common means of polarity identification, along with typical constructions and BS symbol are shown in Figure 8.9. On the BS symbol, the negative end is denoted by the solid black bar.

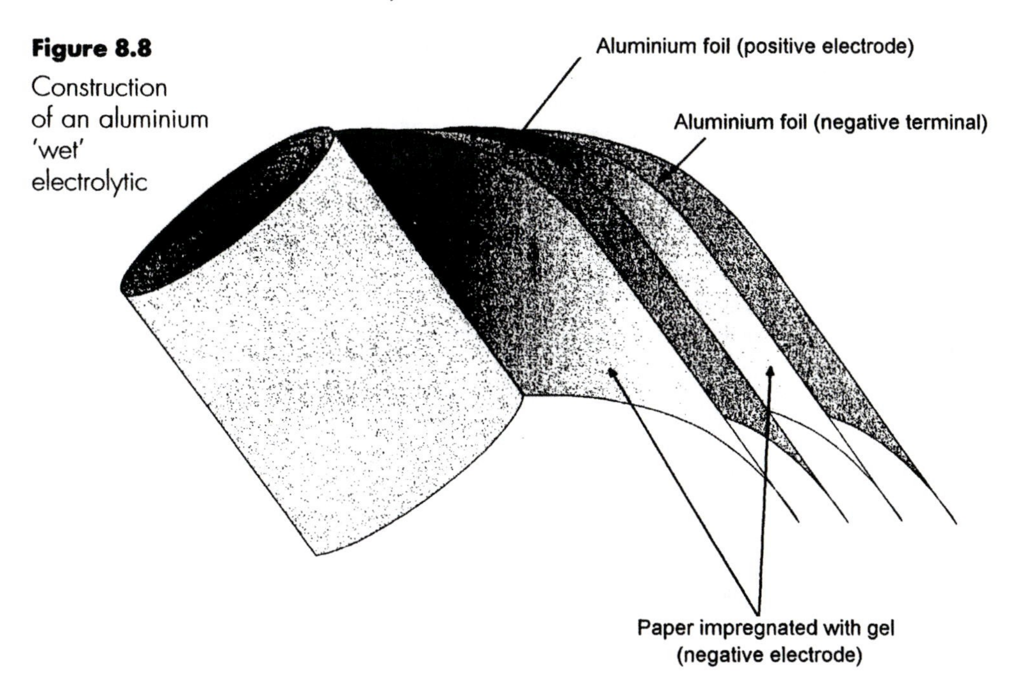

Figure 8.8 Construction of an aluminium 'wet' electrolytic

The physical size of electrolytic capacitors is still quite large, especially for higher voltage rated devices. An alternative device for values in the region of 0.1 μF to 100 μF is the *tantalum* capacitor. For equivalent value electrolytic and tantalum devices, the tantalum is considerably smaller physically. In addition the dc resistance of tantalum is much higher than the aluminium oxide dielectric layer in electrolytics which means that there is a much smaller dc leakage through the capacitor, a factor that is critical in some electronic circuits.

The disadvantaged of tantalum capacitors are that they are more expensive than their electrolytic counterparts, are not available in high working voltages, and have a range of values limited to around 100 μF maximum.

Tantalum capacitors are polarity conscious and, like wet electrolytic capacitors, must be fitted the correct way around. The BS symbol is the same as that for electrolytics shown in Figure 8.9.

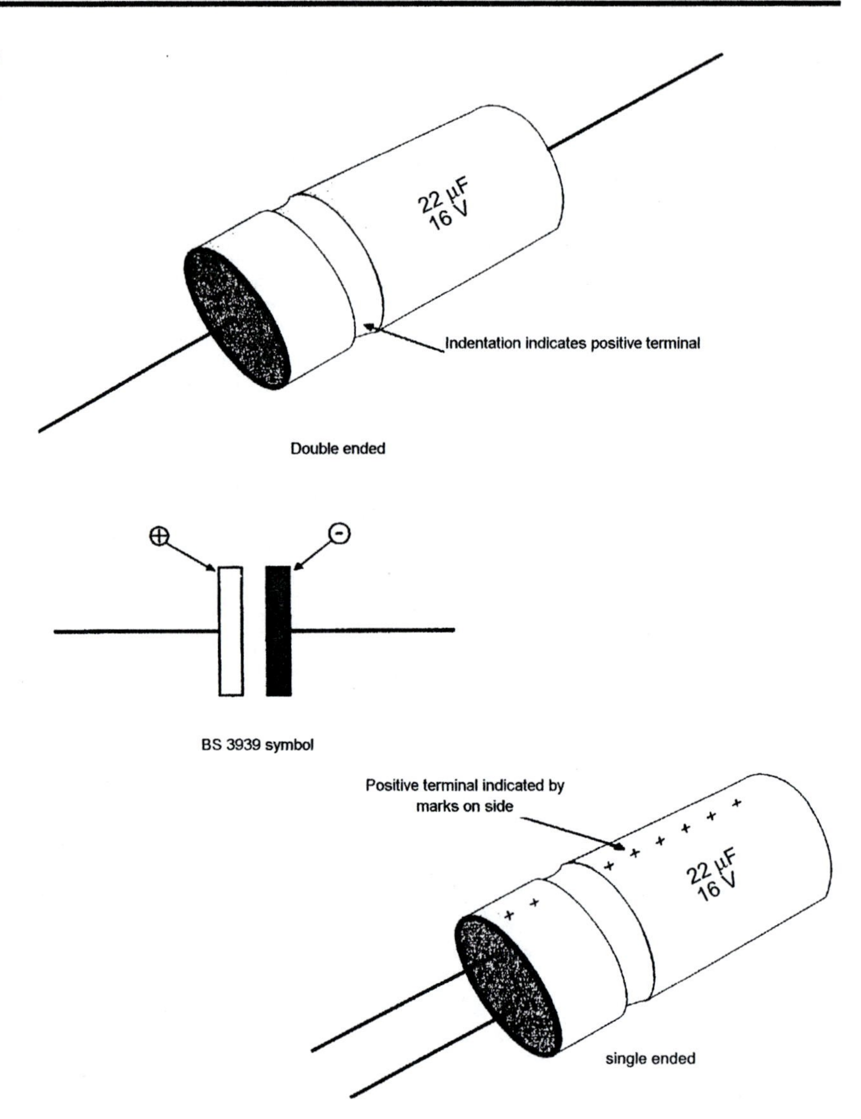

Figure 8.9
Electrolytic capacitors

Air spaced capacitors are constructed from two sets of parallel plates separated by a gap of around 0.25 mm–1 mm. One set of plates is fixed, while the other set is mounted on a shaft enabling them to move in and out of the fixed set. This movement alters the effective area of plate in the capacitor thus making it a *variable capacitor.* Figure 8.10 shows typical types, along with the BS symbols.

Variable capacitors are limited to just a few tens of picofarads in value, but this value is sufficient for tuning radio rf and other high frequency circuits.

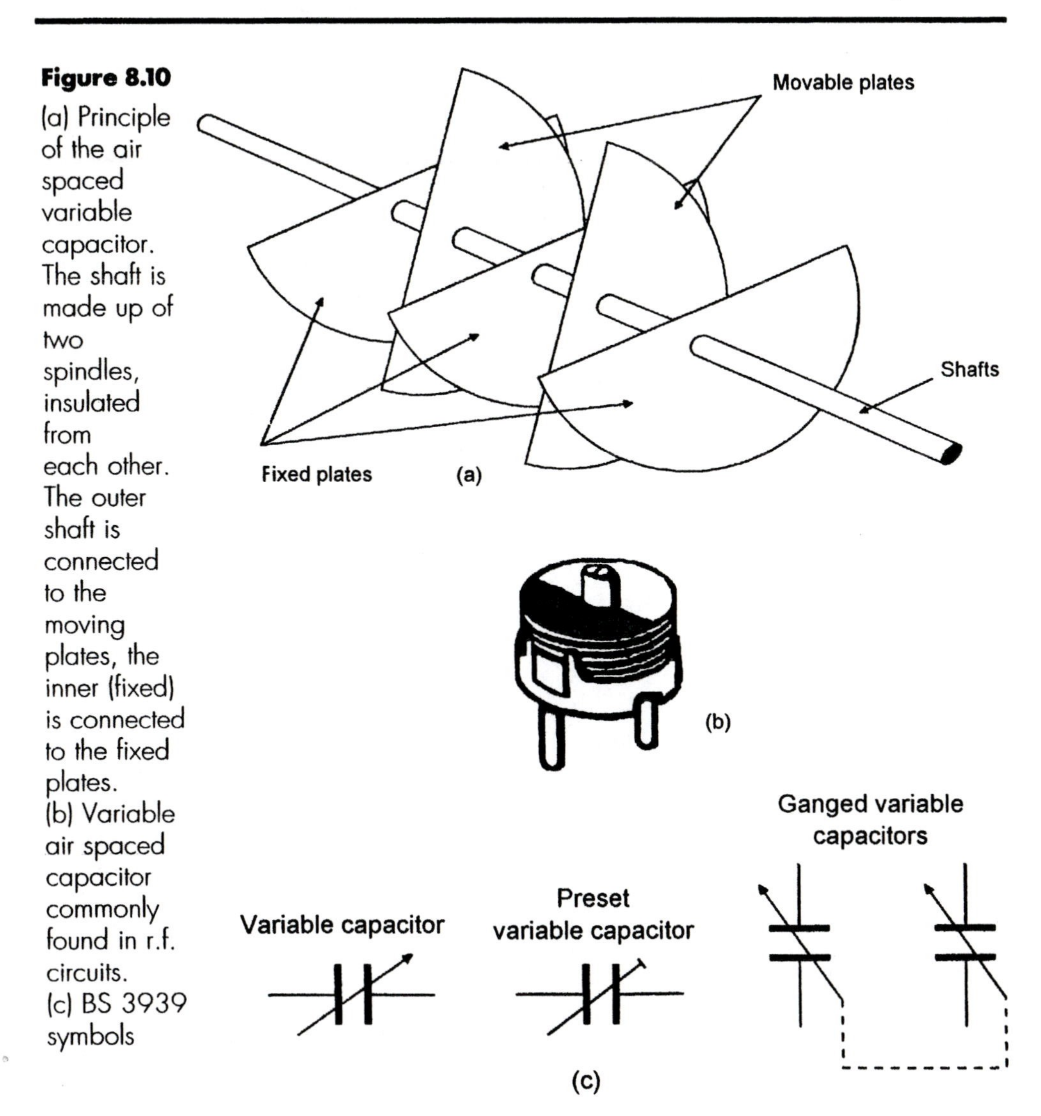

Figure 8.10 (a) Principle of the air spaced variable capacitor. The shaft is made up of two spindles, insulated from each other. The outer shaft is connected to the moving plates, the inner (fixed) is connected to the fixed plates. (b) Variable air spaced capacitor commonly found in r.f. circuits. (c) BS 3939 symbols

Capacitor coding

There are different methods of identifying the value of capacitors. In some instances the value (and possibly voltage rating) is printed onto the side of the device. In other cases a colour code similar to that for resistors is used. Unfortunately there is more than one method of applying the colour code to capacitors which makes it more difficult to recall, although the colours do still represent the same numerical values as for resistors.

Some common examples of code applications are given in Table 8.1.

Table 8.1 Polyester capacitors (picofarads)

Colour	*1st Band*	*2nd Band*	*3rd Band Multiplier*	*4th Band Tolerance*	*5th Band*
Black	–	0	1	20%	–
Brown	1	1	10	–	100 V
Red	2	2	100	–	250 V
Orange	3	3	1000	–	–
Yellow	4	4	10 000	–	400 V
Green	5	5	100 000	5%	–
Blue	6	6	1 000 000	–	–
Violet	7	7	0.01	–	–
Grey	8	8	0.001	–	–
White	9	9	–	10%	–

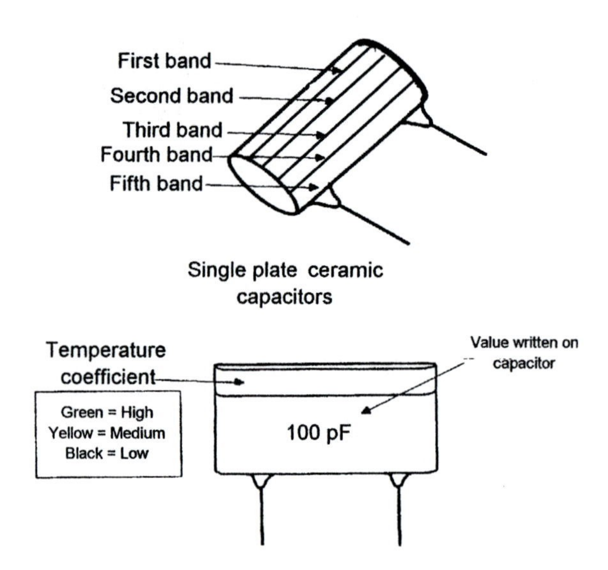

Capacitors in parallel

A circuit showing two capacitors connected in parallel is given in Figure 8.11. Because the value of capacitance is dependent upon the area of the plates, when capacitors are connected in parallel the plate area is effectively increased, so *the total value is the addition of all the capacitors.* The e.m.f. across each capacitor is the same, but because $Q = V \times C$, for parallel capacitors of *differing values* the charge on each capacitor will be different.

Figure 8.11
Parallel plate capacitor

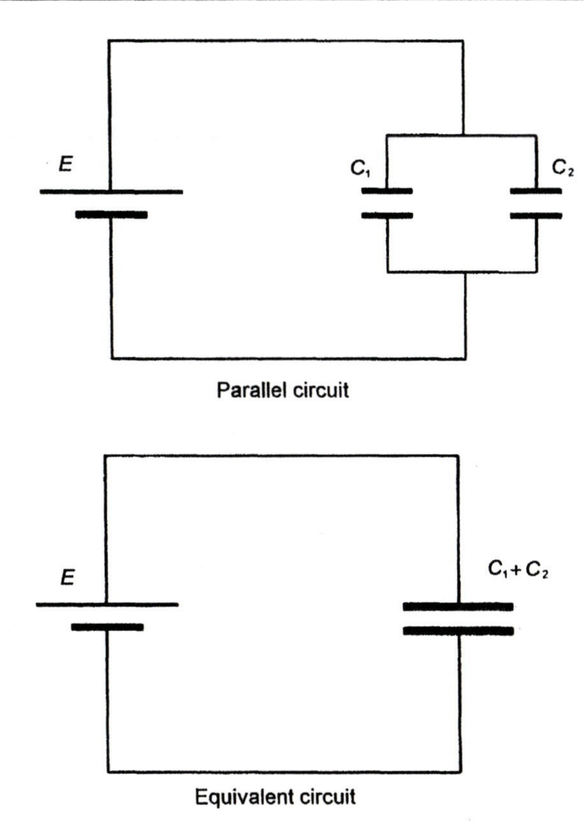

Example

Three capacitors are connected in parallel. If $C_1 = 10\ \mu F$, $C_2 = 50\ \mu F$, and $C_3 = 100\ \mu F$, what is the total circuit capacitance?

Solution

$C_{total} = C_1 + C_2 + C_3 = 10\ \mu F + 50\ \mu F + 100\ \mu F = 160\ \mu F$

In this case the addition was simple because the values were all in the same units, i.e. μF. When the values are in different units you must first convert them into the same units before addition.

Example

Three capacitors are connected in parallel. If $C_1 = 1\ \mu F$, $C_2 = 100$ nF, and $C_3 = 10\ 000$ pF, what is the total circuit capacitance?

Solution

First convert all values to microfarads.

$C_2 = 0.1\ \mu F \quad C_3 = 0.01\ \mu F$

Thus $C_{total} = 1\ \mu F + 0.1\ \mu F + 0.01\ \mu F = 1.11\ \mu F$

We can see that the total value of a parallel capacitor circuit is resolved in the same way as a series resistor circuit.

Although it is rare to actually make capacitor values up by using a number of components (it would, to say the least, appear untidy!), the principle of parallel connection is applied frequently in air spaced construction where the total value is the sum of the capacitance between each pair of plates (Figure 8.12).

Figure 8.12 Parallel construction multiplate capacitor

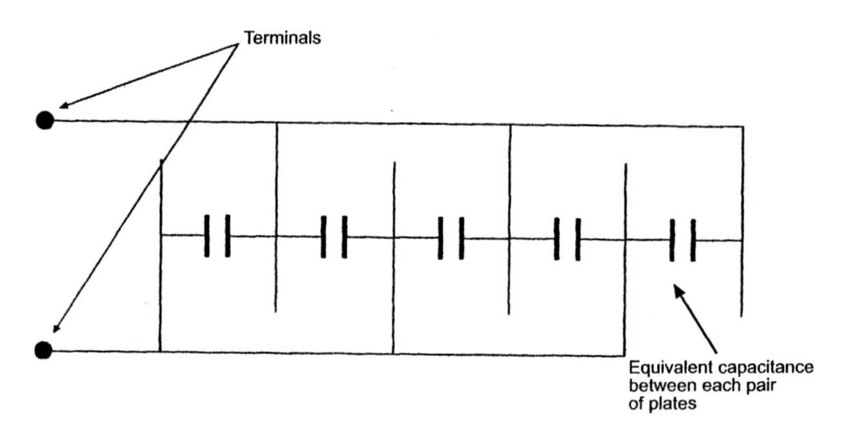

Capacitors in series

The behaviour of capacitors in series tends to be opposite to the parallel arrangement. If we examine the circuit in Figure 8.13 we see that during the charging period electrons are attracted from the left-hand plate of C_1 by the positive pole of the battery (a). The positive charge on C_1 left-hand plate causes an equal number of electrons to move from C_2 to C_1 (b). Finally, an equal number of electrons move from the battery negative to C_2 right-hand plate to satisfy the charge (c). Because an equal number of electrons moved around the circuit, *the charge on each capacitor must be the same.*

Figure 8.13 Movement of electrons through a capacitive circuit

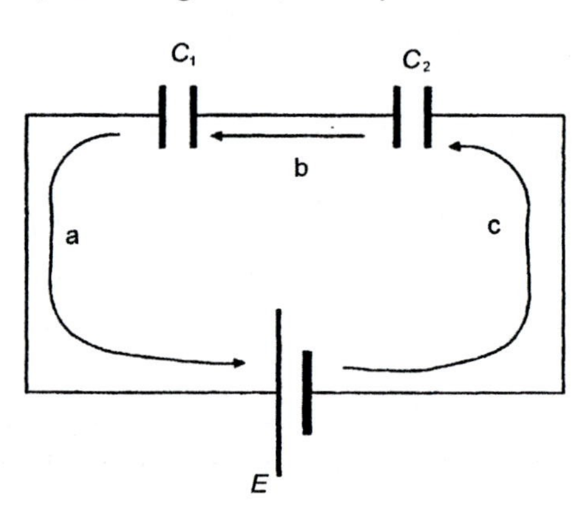

If the values of C_1 and C_2 are different, then from $Q = V \times C$ we see that *the voltage across each is different,* the higher voltage being on the smaller value capacitor.

Example

If $C_1 = 10\ \mu F$ and $C_2 = 100\ \mu F$, then from $Q = V \times C$:

$$VC_1 = \frac{Q}{C_1} \quad \text{and} \quad VC_2 = \frac{Q}{C_2}$$

and because the charge (Q) on each capacitor is the same

$$VC_1 = \frac{Q}{10\ \mu F} \quad \text{and} \quad VC_2 = \frac{Q}{100\ \mu F}$$

by proportion we see that the voltage on C_1 will be ten times greater than that on C_2.

The value of capacitors in series is calculated in the same way as resistors in parallel. The reason for this can be seen when we examine the arrangement shown in Figure 8.14 where two *identical* capacitors are connected in series.

In this arrangement, the effective plates in the circuit are the two outer plates, the inner plates are simply connected together and have no effect on the rest of the circuit. Thus the equivalent capacitance is said to be made up of the two outer plates. The electrostatic field in the equivalent capacitance must pass through a distance equal to the thickness (d) of the two dielectrics, meaning that its strength is reduced. Because the value of a capacitor is proportional to a/d, the value of the total series capacitance is lower than the value of either of the two originals.

When *two* capacitors of different values are connected in series, the total capacitance can be found from

$$C_{total} = \frac{C_1 \times C_2}{C_1 + C_2}$$

Note that when two capacitors of the *same value* are connected in series, the total capacitance will be *half the original value.*

Where more than two capacitors are connected in series, the value is found from

$$\frac{1}{C_{total}} = \frac{1}{C_1} + \frac{1}{C_2} + \frac{1}{C_3}$$

Combination circuits

By applying the above theory for series and parallel capacitors, it is possible to analyse circuits containing a combination of series and parallel connected capacitors. The approach is similar to resistive networks where parts of the circuit are resolved by reducing the combinations until the whole circuit has been resolved.

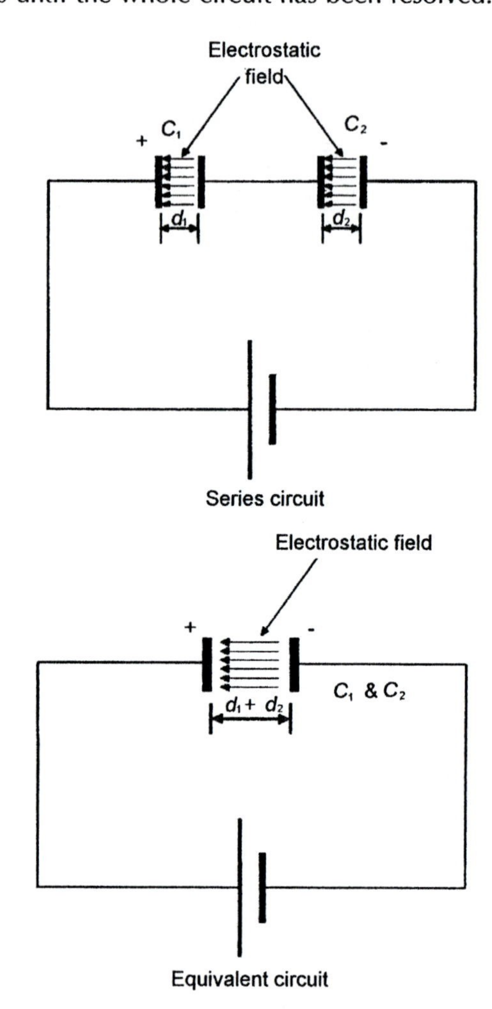

Figure 8.14
Series capacitor circuit

Example

For the circuit in Figure 8.15, find the total capacitance when $C_1 = 10\ \mu F$, $C_2 = 100\ \mu F$, and $C_3 = 100\ \mu F$.

Solution

First resolve the parallel network C_2/C_3, which is the addition of the two values. Thus $C_{23} = 200\ \mu F$.

Next resolve the series circuit containing C_1 and C_{23}. Thus

$$C_{total} = \frac{C_1 \times C_{23}}{C_1 + C_{23}} = \frac{10 \times 200}{10 + 200} = \frac{2000}{210} = 9.5\ \mu F$$

Figure 8.15

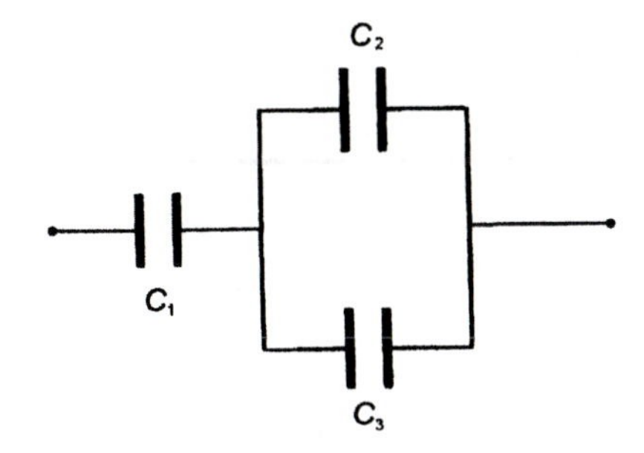

Example

For the circuit in Figure 8.16 calculate: (a) the total capacitance; (b) the charge on C_1.

Solution

(a) $C_{234} = 1/C_1 + 1/C_2 + 1/C_3$

$= 0.05 + 0.01 + 0.02$

$= 0.08$

thus $C_{234} = 1/0.08 = 12.5\ \mu F$

$C_{total} = C_1 + C_{234}$

$= 100 + 12.5 = 112.5\ \mu F$

(b) Charge in C_1 is found from $Q = V \times C$, thus

$Q = 300 \times 100 \times 10^{-6}$

$= 0.03$ coulombs

Figure 8.16

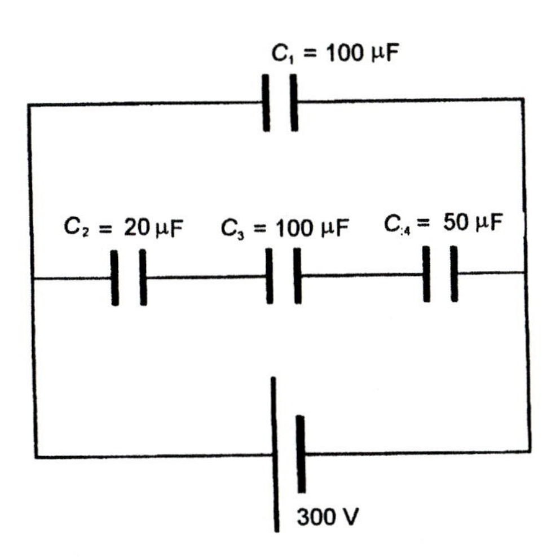

Time constant

This refers to the time that it takes for a capacitor to charge or discharge by a certain amount. In the circuit in Figure 8.17(a), as soon as the switch is closed a high current flows because there is virtually no resistance in the circuit. This high current means that the capacitor will charge very quickly. In Figure 8.17(b) the presence of resistance in the circuit reduces the current, causing the capacitor to take longer to acquire a charge. In Figure 8.17(c), a fully charged capacitor is connected across a resistor, so the rate of discharge when the switch is closed will be reduced, extending the discharge time.

Figure 8.17

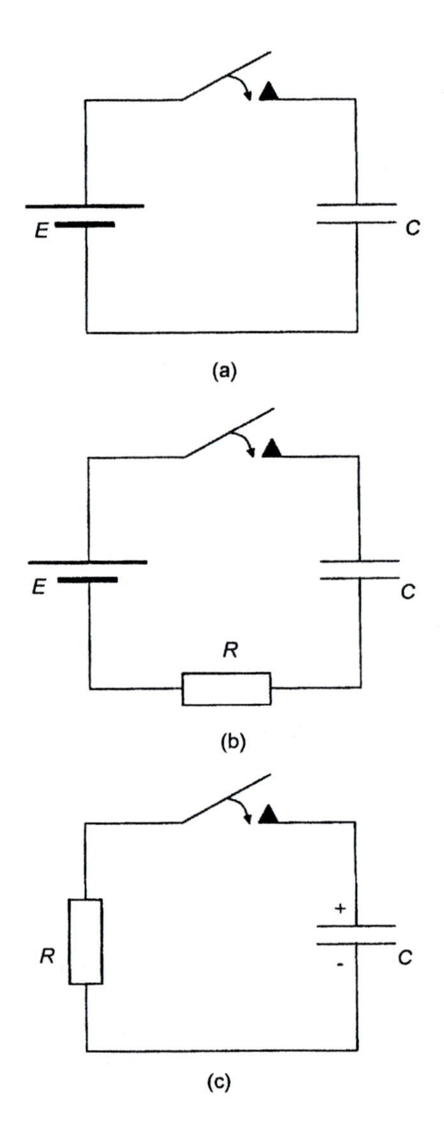

In such a circuit, the time that it takes for the capacitor to charge is determined by the values of both the capacitor and the resistor. The higher the values, the longer will be the charging time.

The *time constant* is the time that it takes for a capacitor to charge to, or discharge by, 63% of the total possible charge, and is found from the expression:

$t = C.R$

where t = time, C = value of capacitor and R = value of resistor.

This equation is derived from the two equations relating to charge, i.e.

$Q = V.C$ and $Q = I.t$

Thus $V.C = I.t$

Transposing for t: $t = C.\frac{V}{I}$

But $V \div I = R$

Therefore $t = C.R$

The rate at which charge/discharge takes place is shown in Figure 8.18. Note that this is not linear because as the capacitor charges the current reduces. The rate of charge and discharge is said to be *exponential*, the essential point of this characteristic being the fact that it never actually reaches 100% when moving in either direction. This means that a capacitor initially charges or discharges at a rapid rate, slowing down after the first time constant period, and takes four more time constant periods to (not quite) completely charge/discharge.

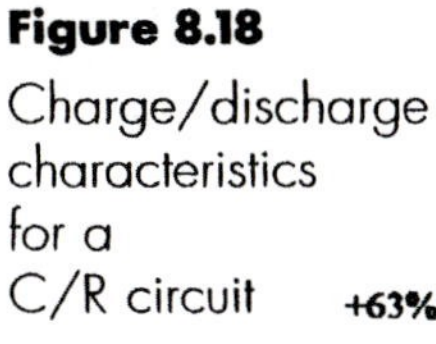

Figure 8.18
Charge/discharge characteristics for a C/R circuit

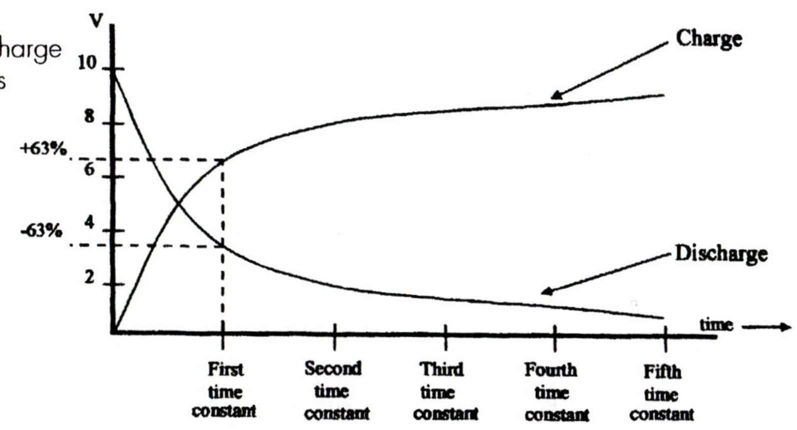

The *C/R* time constant characteristic is exploited to good effect in filter, wave shaping, and timing circuits.

9

Magnetism

Electrical and electronic engineering makes a great deal of use of the property known as magnetism.

What is magnetism?

It is a property possessed by a few naturally occurring materials and some metals, those which are ferrous based and one or two other metals. Not all materials can be magnetised. Examples of materials which can be magnetised are iron, steel, nickel, cobalt and alloys which contain ferrous material.

How can a material be magnetised?

Take a piece of iron and stroke it with one end of a magnet, always stroking in the same direction. It will slowly become magnetised.

A magnet in common use is a compass. After magnetising a pin, suspend it with a piece of cotton and allow it to steady. It will always come to rest in the same direction – pointing north to south. The tip of the pin pointing north is called the north pole whilst the end pointing in the other direction is called the south pole.

In the process of magnetisation the molecules have all been arranged so that they point in the same direction as shown in the bar magnet in Figure 9.1.

Figure 9.1
Bar magnet

If this magnet is now broken into two pieces two separate bar magnets will be formed as in Figure 9.2. Single poles cannot exist on their own so a magnet must always have a north pole and south pole.

Figure 9.2
Bar magnet broken into two pieces

Properties of magnets

If two bar magnets are placed together the effects shown in Figures 9.3 and 9.4 will always occur.

Figure 9.3
Unlike poles attract

Figure 9.4
Like poles repell

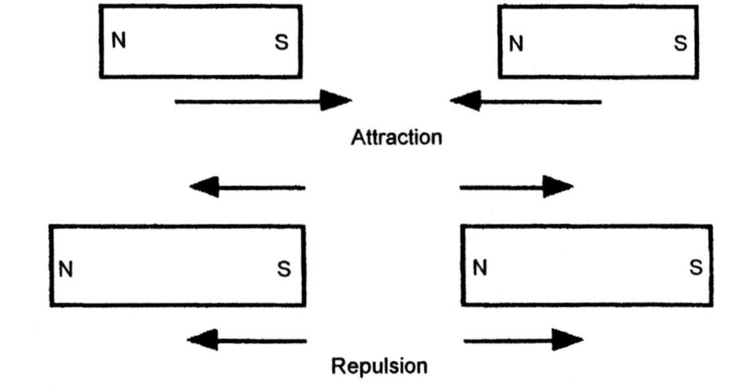

Magnetic fields

Magnetism cannot be seen; however, the effects of a magnetic field can be demonstrated or used. The free space around a bar magnet is under the influence of the magnet and the area where these effects can be detected is called the magnetic field. This magnetic field extends to infinity, but as it moves further away from the magnet its strength reduces and eventually everyday instruments cannot detect it.

It was Faraday who suggested that the magnetic field around the magnet was made up of lines of force. The strength of the magnet being determined by the closeness of the lines of force. The properties of lines of force are always the same in any magnetic field. They are as follows:

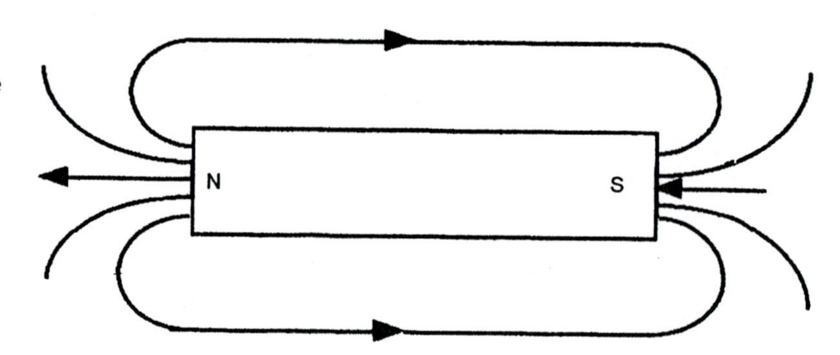

Figure 9.5
Lines of force around a bar magnet

Lines of force are continuous.
Lines of force do not cross.
Lines of force travel from north to south.
Lines of force travelling parallel to each other tend to move apart.
Lines of force travelling towards each other are in opposition and therefore diverge.
Lines of force contract (act similar to an elastic band).

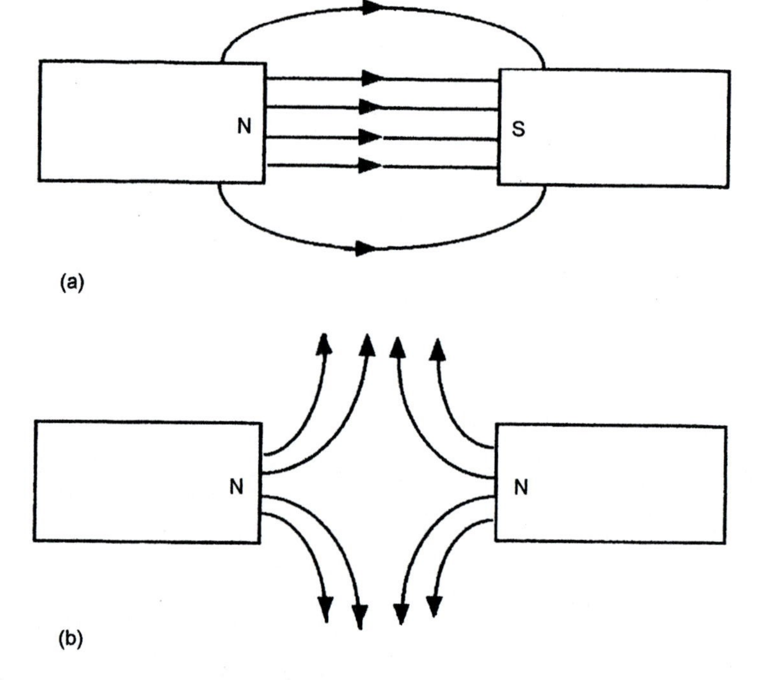

Figure 9.6
(a)Magnetic field between unlike poles.
(b) Magnetic field between like poles

Magnetic field produced by an electric current

Whenever an electric current flows along a conductor it produces a magnetic field around the conductor. The strength and polarity of this magnetic field are determined by the size and direction of the current flow. It is easy to remember the direction of the magnetic field by the use of the corkscrew method which is often referred to as Maxwell's corkscrew rule. If a corkscrew is to be inserted into a cork in a bottle the turning motion is clockwise. This is the same as if the current is flowing into the conductor as shown in Figure 9.7(a) and the magnetic field is in the direction of the turning motion which is clockwise. Removing the corkscrew requires the rotation to be in an anticlockwise direction so the direction of the magnetic field is anticlockwise and is shown in Figure 9.7(b).

Figure 9.7
Direction of field around a current carrying conductor

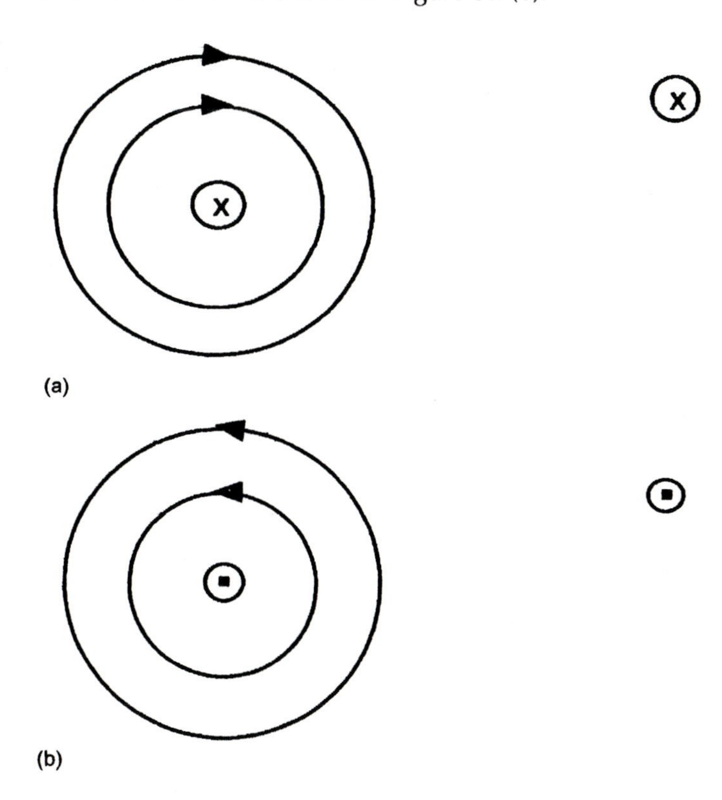

If the current carrying conductor is formed into a coil by winding it around a former the magnetic field produced around each conductor adds together to produce an overall field as shown in Figure 9.8. A coil made like this is sometimes referred to as a choke. Just as in the bar magnet a north–south magnetic field is produced. The strength of the magnetic field depends on the size of the current in the conductor.

It is important to remember that current flowing in a conductor always produces a magnetic field. Consider another case where two conductors are placed side by side and run in parallel. The current is flowing in the same direction in both conductors and since the magnetic fields oppose each other in the middle between the two conductors the overall magnetic field takes up the shape shown in Figure 9.9.

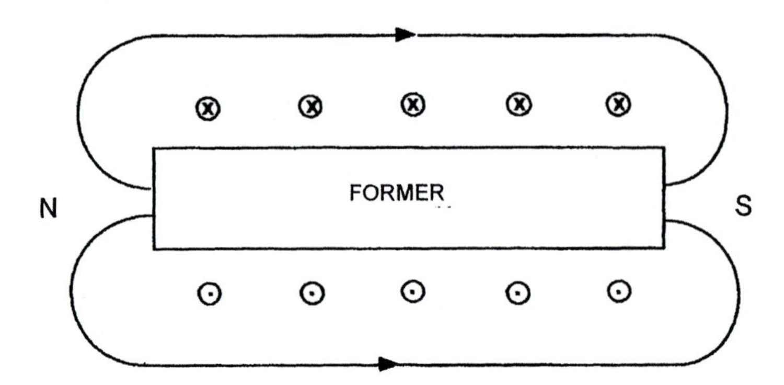

Figure 9.8
Magnetic field around former produced by a direct current

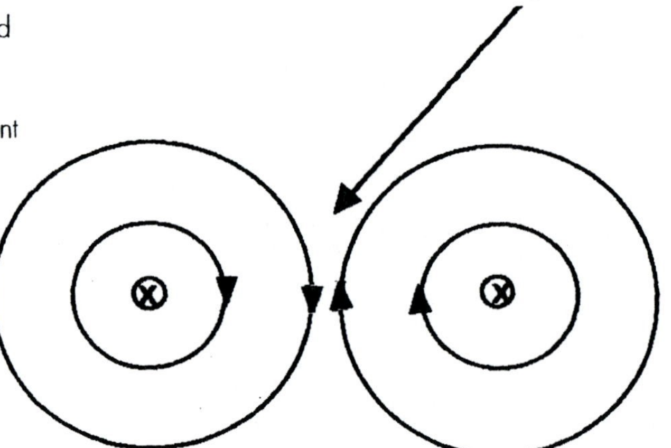

Figure 9.9
Magnetic field around two conductors with the current flowing in the same direction

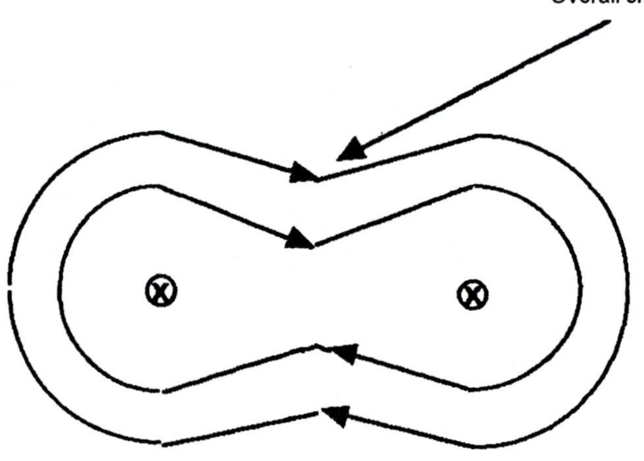

If the current flowing in the two conductors of Figure 9.9 is modified so that the current now flowing is in opposite directions, a different effect will occur as shown in Figure 9.10. Remembering that magnetic lines of force repel each other but in this example are travelling in the same direction in the space between the conductors and also that they cannot cross each other, the result is a magnetic field in between the conductors that is strong with the lines of force very close together. The magnetic force created will try to push the conductors apart. The force acting between the conductors leads to the definition of the unit of current–the ampere as discussed in Chapter 4.

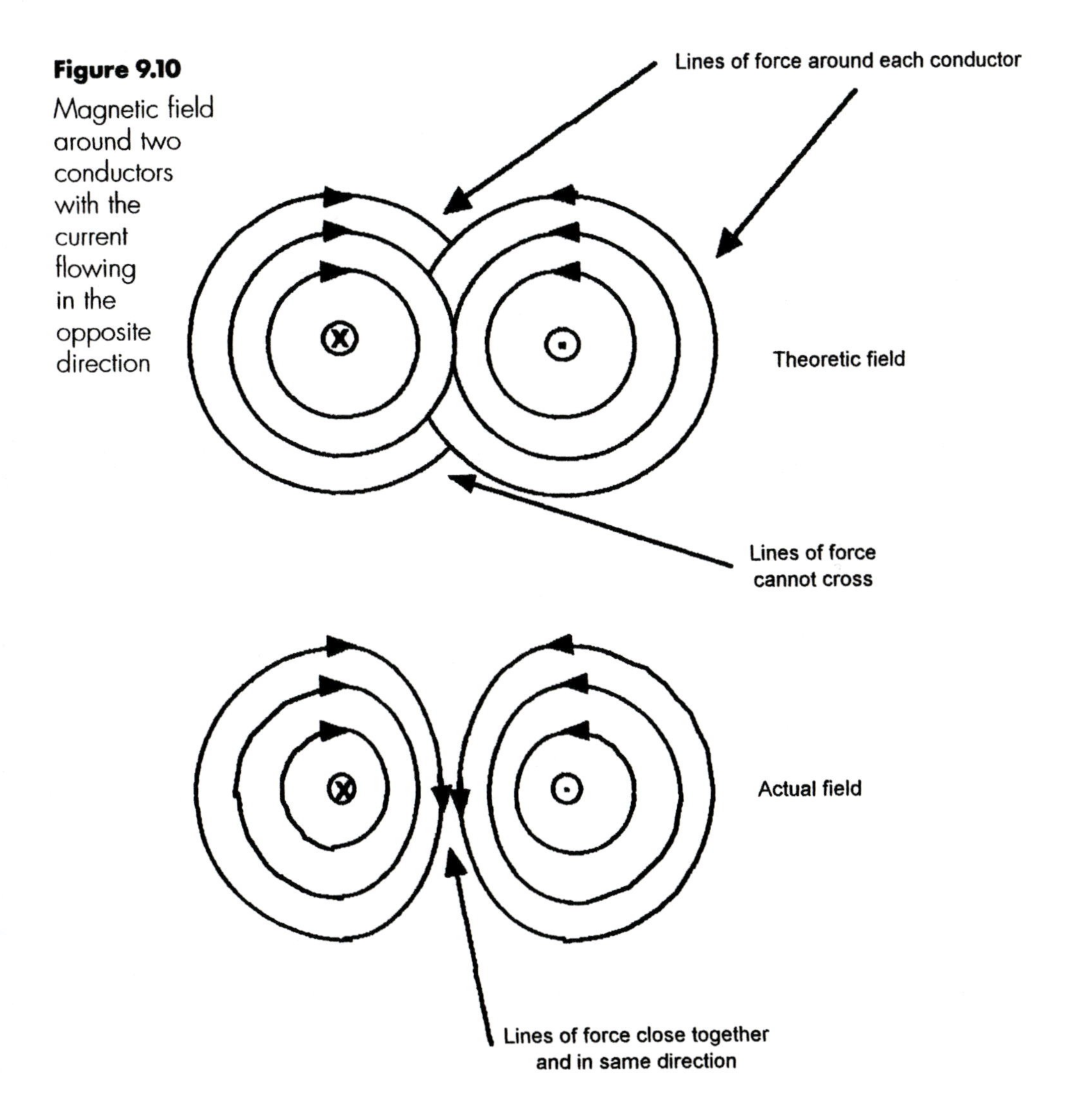

Figure 9.10
Magnetic field around two conductors with the current flowing in the opposite direction

When the current flowing in two infinitely long parallel conductors placed one metre between centres produces a force between the conductors of 2×10^{-7} newtons per metre length of conductor the current flowing is one ampere.

A newton is the amount of force required to move a mass of 1 kg to give it an acceleration of 1 metre per second.

Effect of placing a conductor in a magnetic field

When no current is flowing through the conductor the magnetic field is solely that produced by the permanent magnet and is as shown in Figure 9.11. When current flows through the conductor in the direction indicated by the cross symbol in the centre of the conductor the magnetic field produced is now as shown. This is a result of the interaction between the magnetic field of the permanent magnet and the electromagnetic field produced around the conductor. Because of the interaction of the two fields, the lines of force are close together on the top (fields aiding) and further apart on the bottom (fields opposing), there is a force on the conductor which tries to move it downwards. This is illustrated in Figure 9.12.

Conductor

N

S

Figure 9.11 Magnetic field produced by permanent magnet around a conductor with no current flowing

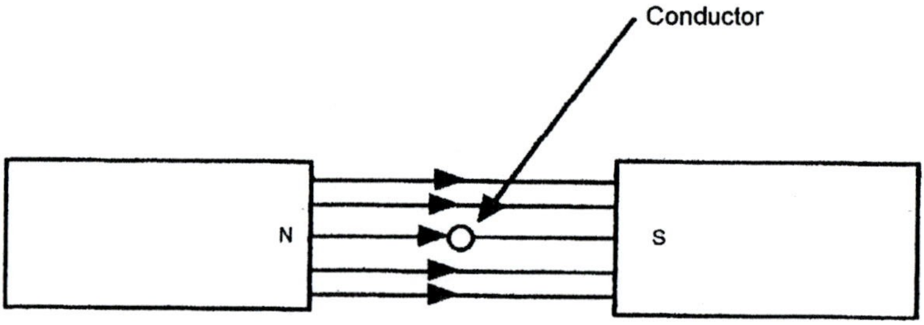

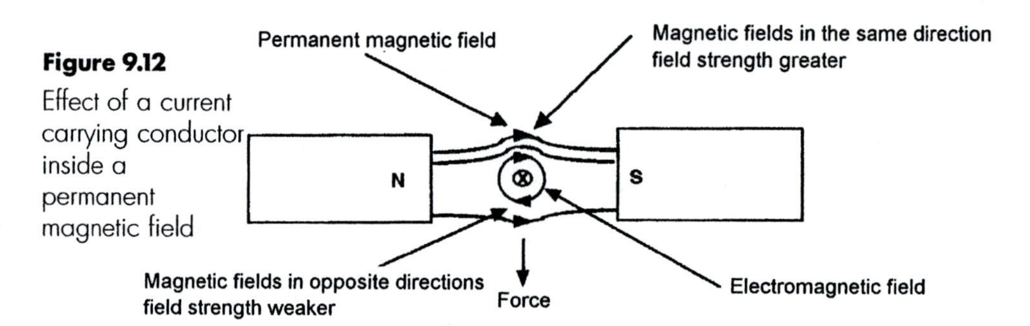

Figure 9.12 Effect of a current carrying conductor inside a permanent magnetic field

If current is passed through the conductor in the opposite direction the resultant magnetic field shown in Figure 9.13 will try to move the conductor upwards.

Examples of where this property is used are the loudspeaker, the electric motor, and the moving coil meter. The total amount of movement or force will depend on:

1. The strength of the magnetic field.
2. The current flowing in the conductor.
3. The length of the conductor in the magnetic field.

Note: The conductor must be at right angles to the magnetic field.
The formula related to this is:
Force (F) = $\beta \times I \times L$
where β = flux density measured in webers per square meter, I = current measured in amps and L = length of the conductor in the magnetic field and at right angles to it.

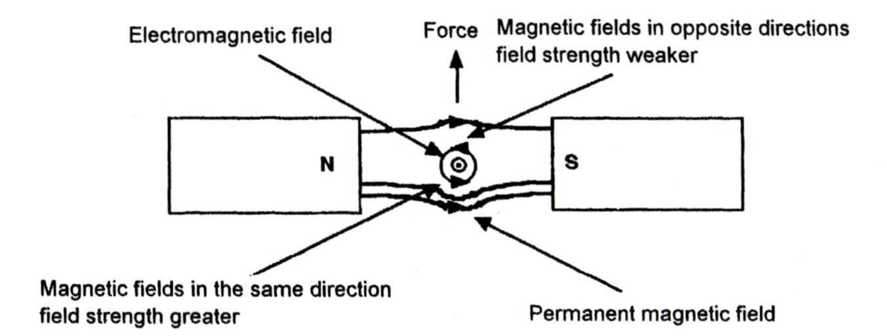

Figure 9.13 Effect of a current carrying conductor inside a permanent magnetic field

Force is measured in *newtons*.

To determine the direction of force on a conductor without drawing the diagrams, Fleming's left-hand rule can be used. The first finger, the index finger, points in the direction of the permanent magnetic field, the second finger is held at right angles to the first finger and points in the direction of the current. The thumb is held at right angles to both fingers and points in the resulting direction of the force/ movement. This effect is illustrated in Figure 9.14, but it is important to remember that this rule applies to conventional current where current flows from positive to negative.

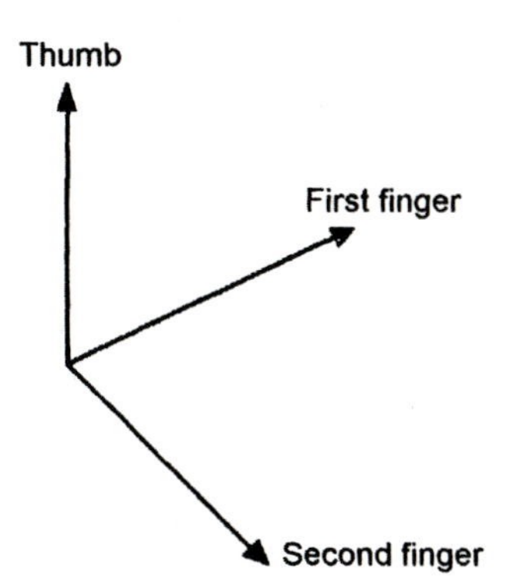

Figure 9.14 Fleming's left-hand rule

Electromagnetic induction

It has already been stated that if a conductor is placed inside a magnetic field and a current is passed along it, the conductor is subjected to a force and may be physically moved.

Force is the result of a magnetic field and current flowing in the conductor.

Suppose that a conductor is placed inside a magnetic field, at right angles to it, and then moved through the magnetic field by applying force. The effect will be to generate an e.m.f. and this phenomenon is called electromagnetic induction. Two factors govern the polarity of the induced e.m.f.: the direction of movement of the conductor through the field and the direction of the permanent magnetic field. This is the principle used in a dynamo or voltage generator. To determine the direction of current and hence the polarity of the voltage produced Fleming's right-hand rule can be applied. This is similar to the previous rule. The first finger points in the direction of the permanent magnetic field, the thumb in the direction of the movement while the second finger shows the direction of the induced current, just as before but it is now the right hand that is used. It is important to note that the two rules identified *cannot* be interchanged. The diagrams shown in Figure 9.15 show the direction of the current produced.

Figure 9.15
The direction of induced voltage/current in a magnetic field

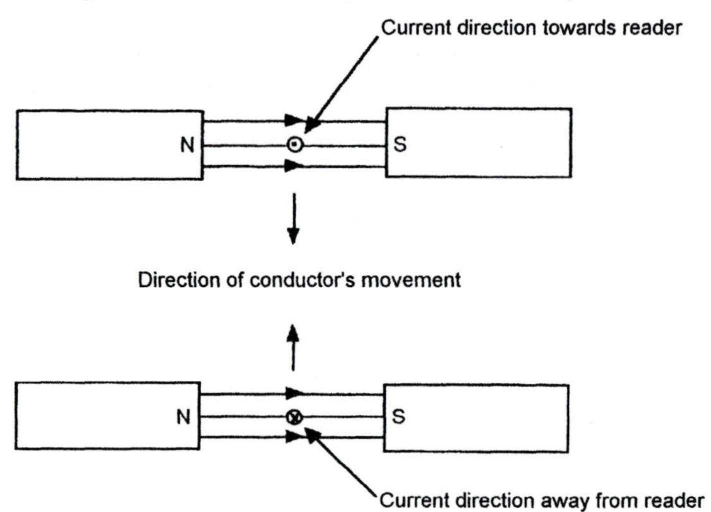

Although the motion which causes the induction of the e.m.f. is upwards, the force reacting against this movement is downwards, or alternatively if the motion is downwards the force reacting against this movement is upwards. This is an example of Lenz's law.

The direction of an induced e.m.f. is always in such a direction that it tends to set up a current opposing the change responsible for inducing the e.m.f.

How large will the induced e.m.f. be? It might be suggested that the electrical power generated is determined by the mechanical power applied to the conductor. Remember the formula quoted previously $F = \beta.I.L$. This can now be amended to take account of the movement and in particular the speed or velocity of movement. It can be established that F, the force, is equal to the electrical power so that:

$F = E \times I$ and $E \times I = \beta \times I \times L$ which can be rewritten as $E = \beta \times L$

Taking account of the velocity V the formula now becomes:

$$E = \beta \times L \times V$$

The induced voltage is determined by the number of lines of force cut per second by the conductor. A conductor of 1 metre moving at a velocity of 1 metre per second across a magnetic field whose strength is 1 weber/m^2 will produce an e.m.f. of 1 volt.

In the preceding example it was the conductor that was made to move through the magnetic field. This movement generated an e.m.f. The amount of e.m.f. generated was determined by the speed at which the conductor moved through the magnetic field's lines of force. Keeping that idea in mind consider a further example.

A conductor is held in a fixed position. A magnetic field is made to cut through the conductor. The lines of force cut through the conductor so an e.m.f. is produced. The facts therefore tell us that to generate an e.m.f.:

The conductor may be moved inside a magnetic field.
The magnetic field may be moved through the conductor.

or

The magnetic field and the conductor can move relative to one another.

In any of these cases an e.m.f. will be produced but only if there is some movement. If there is no movement there will be no e.m.f.

E.m.f. produced by a changing flux

In all the cases considered so far the magnetic field strength has remained constant because a permanent magnet has been used. The strength of a magnetic field may not be constant and can vary. If the magnetic field is created by a current flowing in a conductor then changing the amplitude of the current will change the strength of the magnetic field.

Consider the sine wave shown in Figure 9.16. This shows the rise and fall of current in a conductor through a full cycle. At point A the current is zero so the magnetic field produced by the current must be zero. At point B the current is at a maximum positive value so the strength of the magnetic field is maximum.

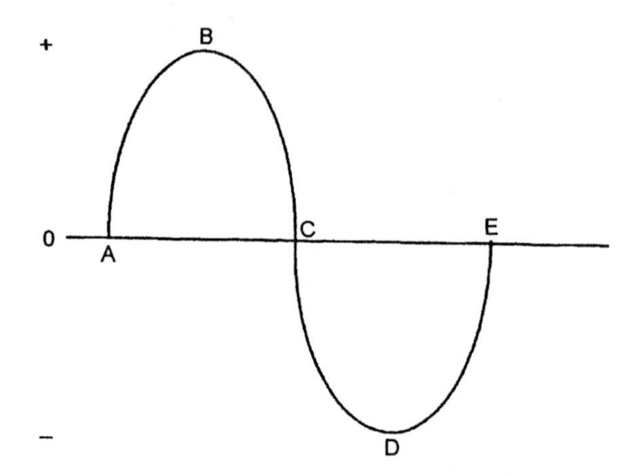

Figure 9.16 Sinusoidal current

Between points A and B the strength of the magnetic field is rising. Between points B and C the current is reducing so the magnetic field produced is decaying. At point C the magnetic field has completely collapsed because the current is zero. In the period between points C and D the current is rising in the negative direction so again the magnetic field gradually increases to a maximum at point D. However, the current is rising in the opposite direction so the magnetic field produced must be in the opposite direction, i.e. it has the opposite polarity so its lines of force travel in the opposite direction. Again between the points D and E the current is falling and eventually at point E is again zero where the magnetic field strength has fallen to zero.

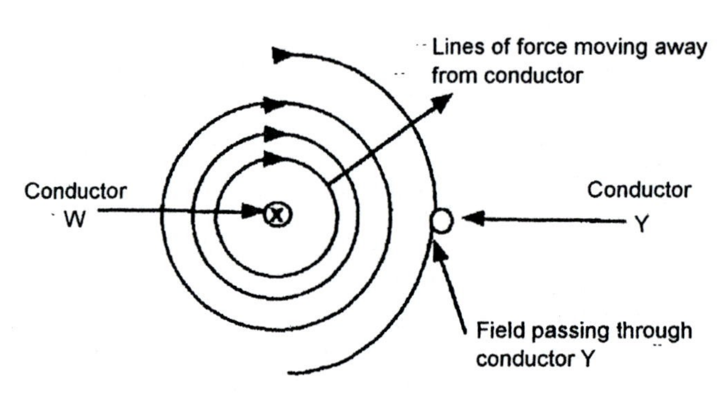

Figure 9.17 Magnetic field passing through conductor with current rising

If the current flowing in conductor W is in the direction indicated a clockwise magnetic field will be produced. As the current is rising towards point B the magnetic field produced is increasing and moving away from the conductor W. By the time the current has reached point B the lines of force produced around conductor W have also cut through conductor Y. This produces a voltage in conductor Y. When the current in conductor W starts to fall, between points B and C the lines of force around conductor W shrink and in doing so again cut through conductor Y producing a voltage of the opposite polarity. This is shown in Figure 9.18.

Figure 9.18
Magnetic field strength reducing with current falling

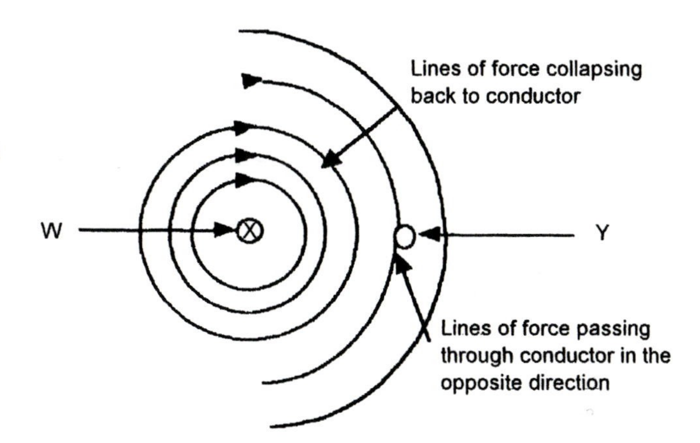

Remember a voltage will only be produced in conductor Y if the magnetic lines of force from the field created by current in conductor W cut through it and that conductors W and Y are parallel to each other.

The magnetic field produced around a single conductor is only very small and as a result the voltage produced in the conductor B is only very small. The effect of producing a voltage in this manner is known by the term mutual inductance. To increase the effectiveness of the mutual inductance effect the conductors are wound into coils and mounted on the same former. The former is made from a magnetisable material, iron (more will be said about this later on).

Figure 9.19
Simple transformer

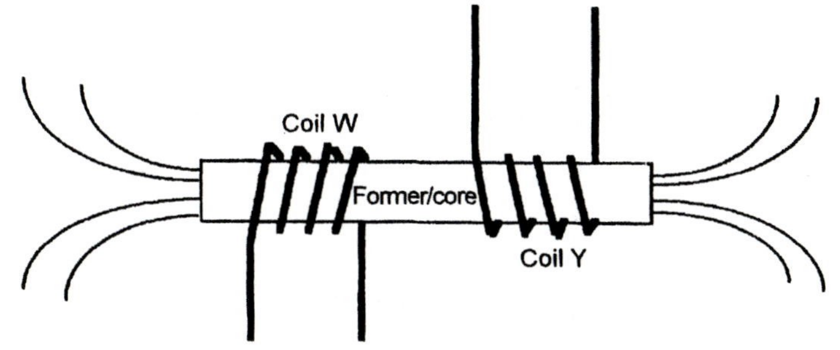

What direction will the induced current in coil Y take? Remember Lenz's law! The flux will set up an e.m.f. in such a direction that the resulting current will try to oppose the current creating it. The current induced will try to stop the field increasing or stop the field decreasing. Hence the current in Y is opposite to W.

When two circuits are arranged so that an e.m.f. is induced in one when the current in the other is changed is commonly known as the transformer effect. Coil W is called the primary and coil Y the secondary. The unit of mutual inductance is the henry.

Core materials

The strength of the electromagnetic field around the conductor is called the magneto motive force (m.m.f.) and is directly proportional to the size of the current and the number of turns of wire used to make the coil. If the coil is free formed, that is, uses air as its core, the relationship between the m.m.f. and the the flux density is linear; however, if the coil is wound around a ferrous core material the flux density is much greater until the point where the core reaches saturation and is magnetised as much as it can be. From the service engineer's point of view this explains how turning a variable core changes the circuit operation since it changes the coil's value of inductance.

Rate of change

The magnitude of the voltage in the secondary will depend upon the rate of change of current in the primary. If the current in the primary changes at the rate of 1 amp/second the induced e.m.f. in the secondary will be 1 volt. This is known as transformer action.

Transformers

The transformer relies on the properties outlined so far. It is not strictly a passive component – like resistors/capacitors – since it is possible to change input voltages/currents to larger output voltages/currents just as an amplifier does. Figure 9.20(a), (b) and (c) show the symbols used in circuit diagrams for transformers. There are different symbols to indicate the type of core material used.

Figure 9.20 Circuit symbols for transformers

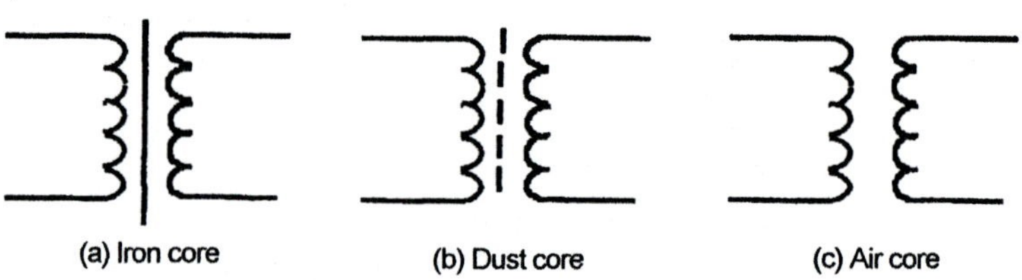

(a) Iron core (b) Dust core (c) Air core

Transformers use the property that a *changing* magnetic field passing through a conductor induces a voltage in it. Examine the diagram of Figure 9.21. The two coils N_p and N_s are linked by a magnetic field produced by the application of V_p. The transformer has a number of turns in the primary winding N_p and a number of turns in the secondary N_s.

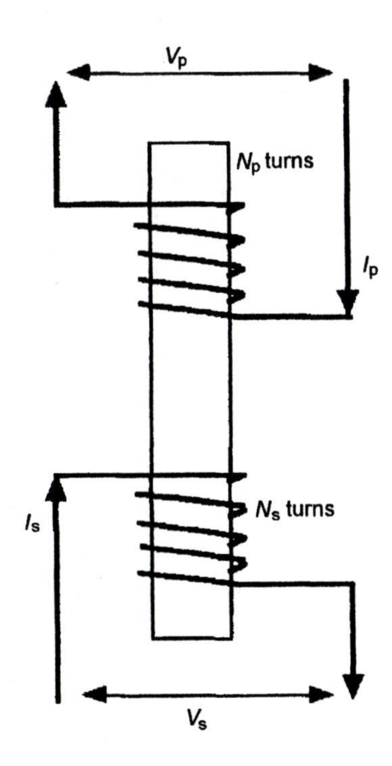

Figure 9.21
Simple transformer

The relationship between the applied voltage and the induced voltage in the secondary is linked by the equation:

$$\frac{V_s}{V_p} = \frac{N_s}{N_p}$$

Suppose the transformer has 150 turns in its primary winding and 150 turns in the secondary winding. When the voltage across the primary winding is 100 volts the voltage induced in the secondary can be calculated as follows:

$$V_s = V_p\left(\frac{N_s}{N_p}\right) = 100\left(\frac{150}{150}\right) = 100(1) = 100 \text{ V}$$

If this transformer is modified so that the number of turns on the secondary is now 300 turns the induced voltage must change, so using the same formula now reveals the output voltage to be larger than the input voltage. The transformer is now a step-up device:

$$V_s = V_p\left(\frac{N_s}{N_p}\right) = 100\left(\frac{300}{150}\right) = 100(2) = 200 \text{ V}$$

The number of turns on the secondary can also be made less than the primary. In the example the effect of reducing the number of turns on the secondary to 75 but keeping the primary turns the same produces a lower secondary voltage. The induced secondary voltage can be recalculated:

$$V_s = V_p\left(\frac{N_s}{N_p}\right) = 100\left(\frac{75}{150}\right) = 100(0.5) = 50 \text{ V}$$

The induced voltage can be varied by changing the relationship between the ratio of numbers of turns on the primary and secondary windings.

Current

The secondary current I_s is related to the primary current I_p by the formula:

$$\frac{I_s}{I_p} = \frac{N_p}{N_p} \quad \text{so} \quad I_s = I_p\left(\frac{N_p}{N_s}\right)$$

Reconsider the first case

$N_p = 150 \quad N_s = 150$

$$I_s = I_p\left(\times\frac{N_p}{N_s}\right) = I_s = I_p \times 1$$

$I_s = I_p$

Reconsider the second case

$N_p = 150 \quad N_s = 300$

$$I_s = I_p\left(\frac{N_p}{N_s}\right) = I_p\,(0.5)$$

Secondary current is half the primary current.
Secondary voltage is twice the primary voltage.
Reconsider the third case

$N_p = 150 \quad N_s = 75$

$$I_s = I_p\left(\frac{N_p}{N_s}\right) = I_s = I_p \times 2$$

Current has increased (×2) – doubled.

Voltage has decreased $(\frac{1}{2})$ – halved.

A transformer has a primary current of 0.5 amps and a primary voltage of 100 volts. It is a step-up transformer with a two ratio of 1 to 10. What will be the secondary voltage and current?

$$V_s = V_p\left(\frac{N_s}{N_p}\right) = 100\left(\frac{10}{1}\right) = 1000\,V$$

$$I_s = I_p\left(\frac{N_p}{N_s}\right) = 0.5\left(\frac{1}{10}\right) = 0.5 \times 0.1 = 0.05\text{ amps}$$

Consider the power used in the primary

Power = $V \times I = V_p \times I_p = 100 \times 0.5 = 50$ watts

Consider the power used in the secondary

Power = $V_s \times I_s = 1000 \times 0.5 = 50$ watts

Primary power = Secondary power.

This assumes 100% efficiency. The secondary power cannot be greater than the primary power. In practice the primary power is always greater since there will be energy lost within the transformer.

Example

A step-down transformer has a turns ratio of 25:1 ($N_p = 25 \quad N_s = 1$). The input voltage is 750 volts. The power delivered from the secondary winding is 500 watts. How much current is flowing in the primary winding.

$$V_s = V_p\left(\frac{N_s}{N_p}\right) = 750\left(\frac{1}{25}\right) = 750 \times 0.04 = 30\,V$$

Power in secondary = $V_s \times I_s$

$$I_s = \frac{\text{Power in secondary}}{V_s} = \frac{500}{30} = 16.67\text{ amps}$$

Power in primary = Power in secondary = 500 watts

$$I_s = \frac{\text{Power in primary}}{V_p} = \frac{500}{750} = 0.67\text{ amps}$$

$$I_p = \frac{L_s}{\left(\frac{N_p}{N_s}\right)}$$

$$I_p = \frac{16.666667}{25} = 0.666\text{ amps}$$

Transformers can also be manufactured with a number of windings. It is not unusual for an electronic device to need several different voltages to enable it to function, a 12 V supply and a 5 V supply. This can be provided by a single transformer with two secondary windings as shown in Figure 9.22. The ratio of the number of turns required in each secondary winding is determined in exactly the same manner as described previously.

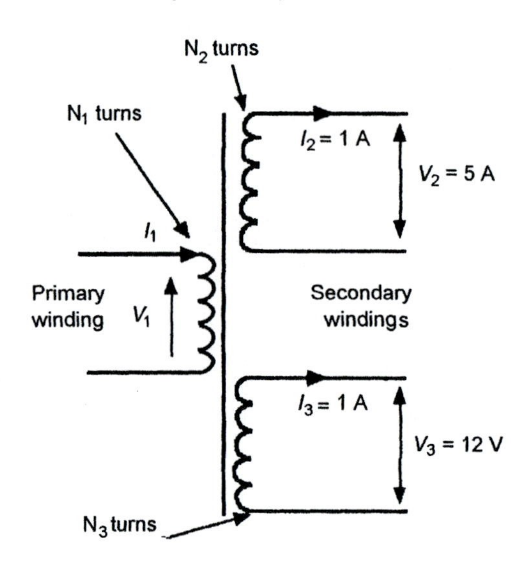

Figure 9.22 A multiple winding transformer

Transformer VA rating

The primary winding of a transformer provides all the energy for the secondary windings. The power in the primary winding must therefore be equal to the sum of the powers of the secondary windings and since power is the product of voltage and current the VA rating for the transformer shown in Figure 9.22 can be determined as follows:

Transformer VA rating

$$
\begin{aligned}
V_p I_p &= V_2 I_2 + V_3 I_3 \\
&= 5 \times 1 + 12 \times 1 = 5 + 12 = 17 \text{ VA}
\end{aligned}
$$

10

Semiconductors

In Chapters 4 and 5 basic structure of atoms and the flow of electric current was explained. Some materials were identified as insulators and others as conductors. A knowledge of the physics of materials is a help when examining how semiconductor devices such as transistors and diodes work. The difference between a conductor's atomic structure and an insulator's atomic structure is the number of electrons in the valence shell. Four or more electrons for an insulator and three or fewer for a conductor. Semiconductors can be described as materials which have conducting properties intermediate between conductors and insulators. The two major semiconductor elements are silicon and germanium although others such as gallium arsenide and lead sulphide are used but for transistors it is mainly silicon that is employed.

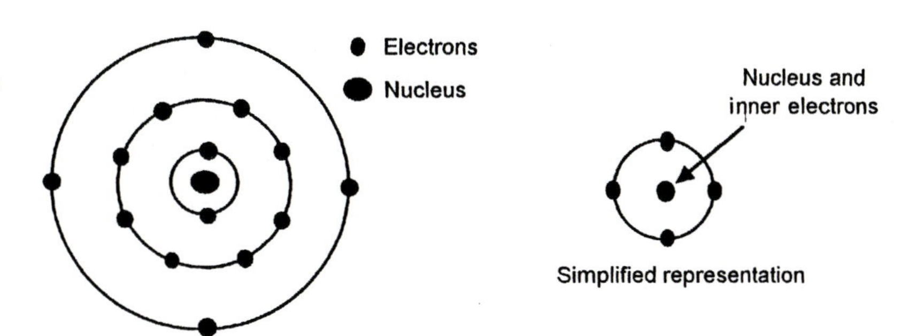

Figure 10.1
Silicon atom

The silicon atom shown in Figure 10.1 has four electrons in its valence shell. From the description given earlier it would appear to fall into the category of an insulator because there are no free electrons present to act as current carriers. When located closely together the valence electrons of adjacent atoms lock together by sharing an orbit. This is referred to as covalent bonding and is shown in Figure 10.2. In practice a material receives energy mainly as heat which excites electrons and can cause them to break out of their orbit. This increases the conductivity of the material so it is no longer a perfect insulator. As an electron moves out of its orbit it creates a *hole,* hence the atom has lost its neutral charge. It is a negative electron short so has become positively charged. It will therefore seek to neutralise itself by capturing an electron from another atom. The hole does not stay static because by stealing an electron from a neighbouring atom a fresh hole appears. This means that current flow depends upon both electrons and holes moving. Free electrons and holes are called current carriers.

The amount of electron/hole movement in a piece of silicon would be very small. To increase its conductivity to usable levels it is *doped.* Doping is a process where a small amount of impurity is added. The impurity added is carefully chosen to give the desired affect. A typical impurity could be arsenic or antimony. Both of these elements are *pentavalent* and have five electrons orbiting in their valence shell. It therefore seems logical that when combined with silicon which only has four valence electrons there will always be a spare electron since it is unable to form a covalent bond with another electron. This is shown in Figure 10.2(b).

Figure 10.2
Doped silicon crystal,
N type material

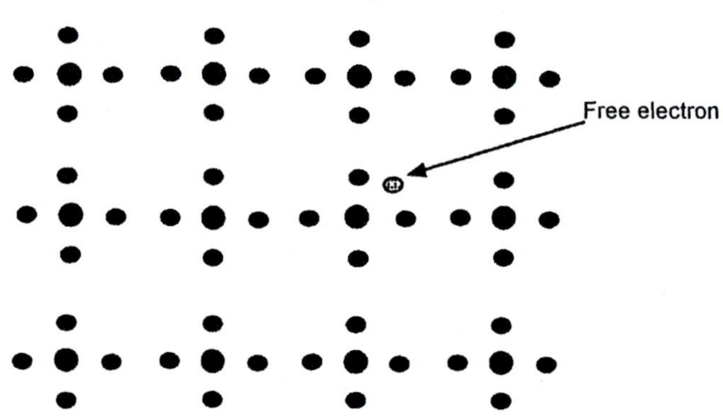

Alternatively a small amount of impurity can be added that has only three valence electrons, typically aluminium or indium. The material now has an electron missing or contains a hole. This is shown in Figure 10.3.

Two different semiconductor materials have been created. The material with spare electrons is called *N type material* while the material with holes is called *P type material.* Silicon is a much commoner element than germanium hence the majority of transistors and other semiconductor devices are manufactured from silicon. The facts outlined so far are true for both silicon and germanium although in operation there are some differences which will be identified.

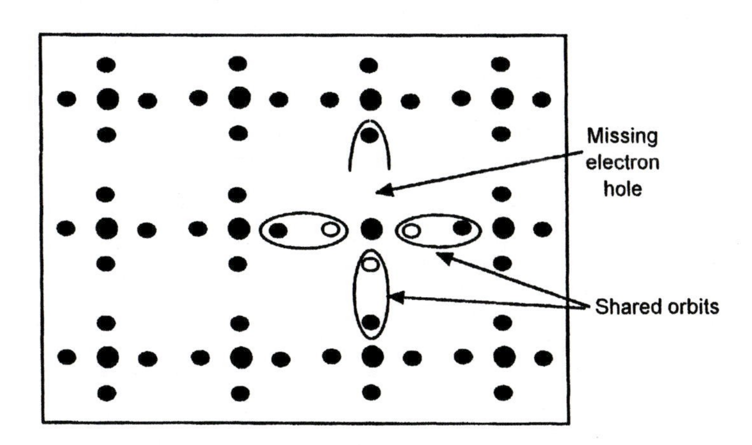

Figure 10.3
Doped silicon crystal, P type semiconductor material

The semiconductor diode

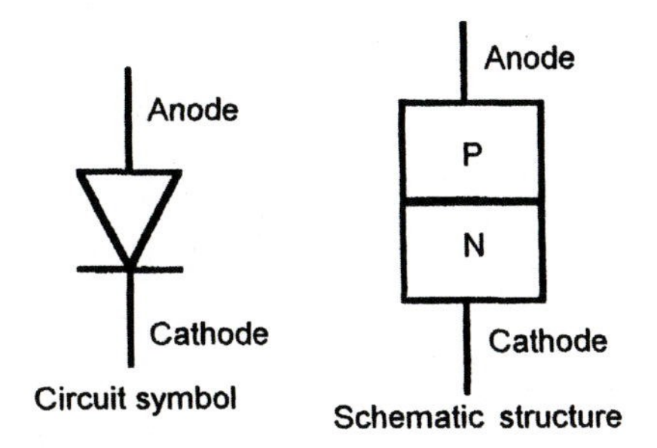

Figure 10.4
The P–N junction diode

Figure 10.4 shows the circuit symbol for a P–N junction diode along with its schematic structure. The device has two electrodes, an anode and a cathode. The anode is the P type material and the cathode the N type material. If the diode is connected to a battery as shown in Figure 10.5, current will flow. The spare electrons in the N type material will be driven away from the negative terminal of the battery since like charges repel. Similarly the holes in the P type material will be driven away from the positive terminal of the battery. If the battery has sufficient force, e.m.f., the electrons in the N type material will be forced into the P type material and since electrons are moving, current flows. The device is said to be *forward biased.* Any electrons leaving the N type material are replaced by electrons from the negative pole of the battery.

Figure 10.5
Forward biased diode

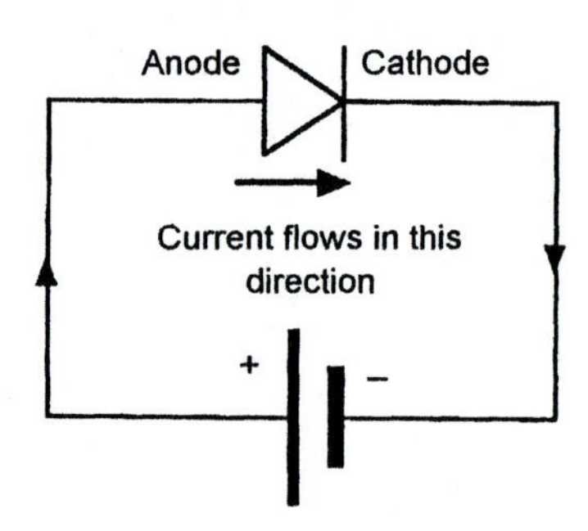

If the battery connections are changed so that they are as connected in Figure 10.6 the diode is now *reverse biased* and no current flows. The electrons in the N type material are attracted to the positive terminal of the battery while the holes are attracted to the negative terminal of the battery. There is now no chance of the free electrons moving into the holes so there is no current flow.

Figure 10.6
Reverse biased diode

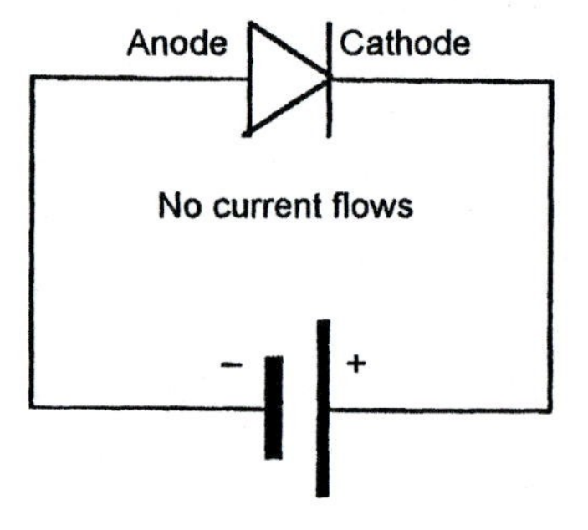

As a result diodes have a low resistance to current flow when forward biased and a high resistance to current flow when reverse biased. A semiconductor junction can be tested using an ohmmeter. It should give a low resistance reading when the meter leads are connected in one direction and a high resistance reading when they are reconnected in the other direction. The important point to remember is that for a diode to conduct its anode must be more positive than its cathode. The phrase often used to describe this condition is *positive with respect to.* How much greater the anode voltage must be depends upon the material used to manufacture the diode. Germanium diodes conduct with a lower potential difference than silicon diodes. The voltage required to turn on a germanium diode is 100 millivolts anode/cathode potential difference whilst for a silicon diode this potential difference must rise to 500 millivolts. The characteristics of a P–N junction diode are shown in Figure 10.6.

Figure 10.6 shows that once the turn-on voltage has been exceeded a very small rise in voltage produces a large rise in current in the forward direction. It is important to note that the current in the diode cannot be allowed to rise beyond the level that the diode has been manufactured to withstand or it may be permanently damaged. In the reverse direction, where the diode is said to be non-conducting, a small leakage current is shown. This leakage current is so small that the scales have to be changed in the diagram to show it. The leakage current appears to be almost constant up until the breakdown point. When the reverse bias potential reaches this point there is a sudden increase in current. This sharp rise in current will almost always lead to the destruction of the diode unless it is a special diode called a *zener diode* which is specifically designed to operate in reverse bias at its breakdown potential. This diode will be dealt with later. When constructing circuits using diodes care must always be taken to ensure that the diode does reach a condition where the amount of reverse bias applied exceeds the manufacturer's safe working limit, known as the *peak inverse voltage* or *P.I.V.*

Diodes have many uses but can be divided into two groups. Signal diodes and rectifier diodes, according to their application. A signal diode usually requires a low forward voltage drop so germanium diodes are suited to this function, e.g. demodulation, while diodes that have to deal with large currents and voltages are generally silicon diodes used for rectification. In either category the diode will only allow current to flow in one direction. A simple application of rectification is shown in Figure 10.7 where full cycles of alternating current are applied to the anode of the diode but only positive half cycles of output are obtained. This can easily be explained by remembering the fact stated earlier that the diode will only conduct when its anode is positive with respect to the cathode.

Figure 10.7 shows the characteristics of a diode. In the forward direction germanium diodes conduct at a lower forward voltage than their silicon counterparts and as a result germanium diodes are used primarily for signal detection, whilst silicon diodes are employed for rectification purposes, for example in a power supply as well as for other general purposes. If a current of 1 mA is allowed to pass through a germanium diode the voltage drop measured across its electrodes would be 0.3 volts, but if the same current were allowed to pass through a silicon diode the voltage drop measured across its electrodes would be 0.65 volts. In the reverse direction very little current will flow. Note that in Figure 10.7 the reverse current is measured in microamps until the breakdown voltage is reached. At this point the current increases very rapidly and the diode junction will be permanently damaged, unless the diode is of a special construction and is a device known as a zener diode.

Figure 10.7
Diode characteristic

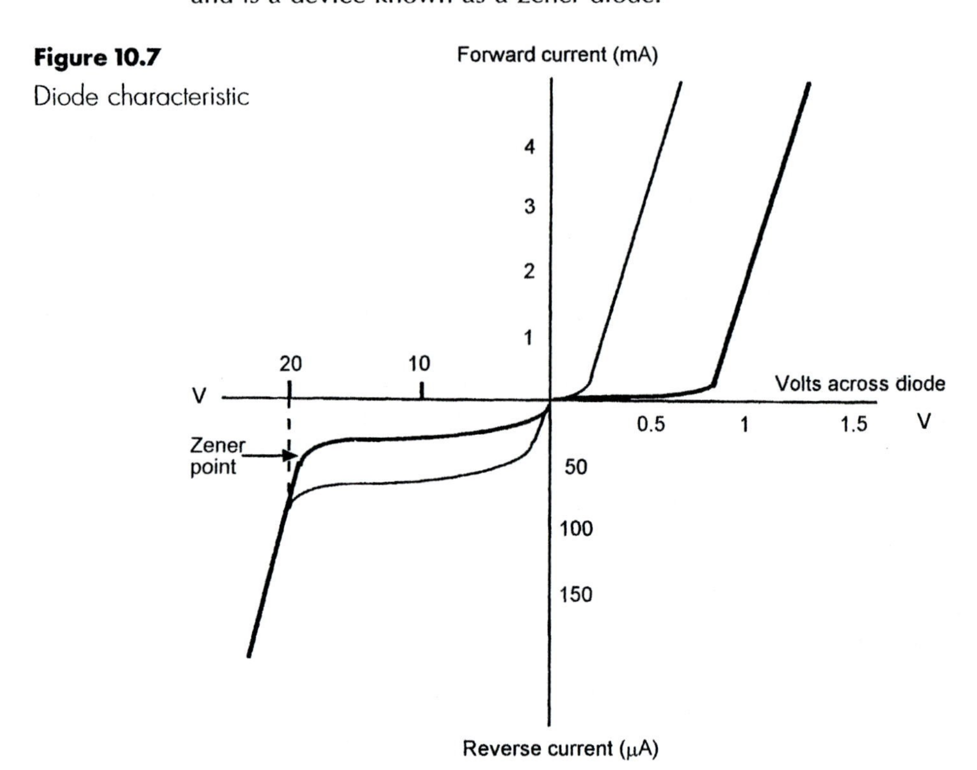

Diode applications

As previously stated a typical application for a diode is that of a rectifier. Figure 10.8 shows a simple rectifier circuit along with its input and output signals.

Figure 10.8
Half wave rectification

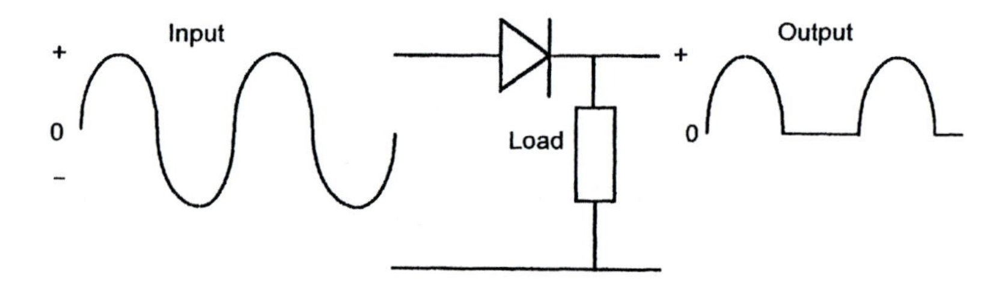

In Figure 10.8 the diode only conducts when its anode is positive and the result is a series of positive half cycles across the load represented by a resistor in this diagram. Normally a smoothing circuit would be added to level out the pulses to produce a steady voltage. Since only half of the input waveform is used, this circuit is known as a half wave rectifier.

Zener diodes

This type of silicon diode is specially designed so that it can be operated under reverse bias conditions and produce a consistent reverse breakdown characteristic. In other words the potential difference between its electrodes will remain constant. This makes it ideal as a voltage reference source. On a circuit diagram the symbol used for this type of diode is a slightly modified version of the normal diode symbol and is shown in Figure 10.9. The physical appearance of a zener diode is the same as that of a conventional diode and just like other diodes can be encapsulated in metal, plastic or glass with the cathode identified by a band, see Figure 10.9(c). When used as a voltage reference source two criteria have to be taken into account to choose the right diode: the reference voltage required and the power dissipation ability of the diode. The circuit of Figure 10.10 shows a zener diode used to provide a voltage reference level.

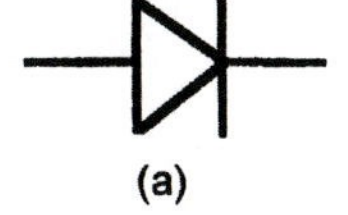

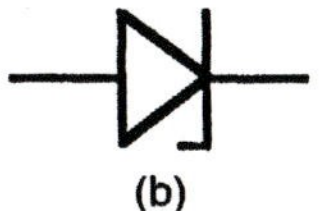

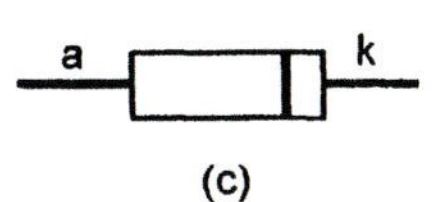

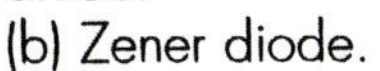

Figure 10.9
(a) Standard diode.
(b) Zener diode.
(c) Diode's appearance

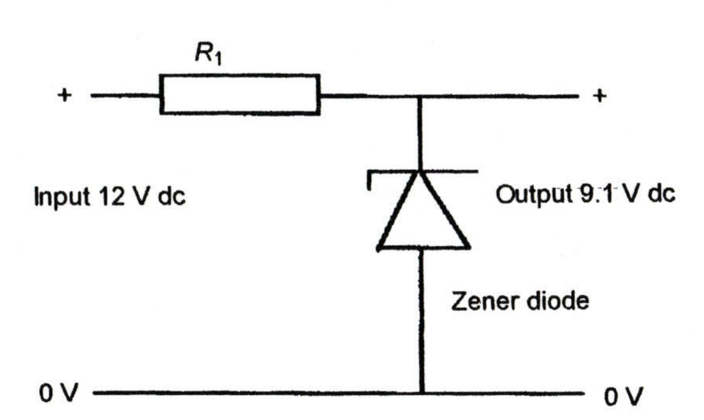

Figure 10.10
Simple zener diode regulator

The operation of this circuit is dependent on the value of the resistor chosen for R_1, the voltage reference and power rating of the zener diode. Suppose a zener diode from the BZX 55 range is chosen. This is a common range of zener diode. The characteristics supplied by the manufacturer reveal that it is a low power device rated at 500 mW and that its zener voltage will fall between 2.4 volts and

91 volts. For the example chosen the diode is a 9.1 volt zener diode. In Chapter 5, covering principles and formulae, the relationship between power, resistance, voltage and current was explained. The ability to manipulate these formulae is now extremely helpful. A device with 9.1 volts across it with 500 mW of power dissipated has a current of 55 mA flowing since current can be determined by dividing the power by the voltage:

$$I = \frac{\text{Power}}{\text{Voltage}} = \frac{500\ \text{mW}}{9.1\ \text{V}} = 54.945\ \text{mA}$$

The resistor R_1 which is in series with the diode must therefore have 54.945 mA flowing through it. If the voltage at the input is 12 volts and there is only 9.1 volts at the output, 2.9 volts have been dropped by R_1. The correct value for this resistor can be calculated by applying Ohm's law:

$$R = \frac{V}{I} = \frac{2.9\ \text{V}}{54.945\ \text{mA}} = 52.8$$

This value represents the worst case where the diode is operating at its maximum dissipation which is far from ideal. If R_1 is increased in value to 470 Ω the voltage dropped across it will still be 2.9 volts because of the action of the zener diode maintaining 9.1 volts across its electrodes. However, the important factor is that the current flowing is now much smaller and the power being dissipated by the diode is also much smaller. Both of these facts can be proved by the application of some simple mathematics. The current flowing is determined by:

$$I = \frac{V}{I} = \frac{2.9\ \text{V}}{470} = 6.17\ \text{mA}$$

$$\text{Power} = V \times I = 9.1\ \text{V} \times 6.17\ \text{mA} = 60.97\ \text{mW}$$

The diode is not having to work quite so hard and this would improve its reliability.

Variable capacitance diodes

This is yet another type of diode. It is operated with its junction reverse biased and the layer formed between the P and N type materials, known as the depletion layer, acts as the dielectric in a capacitor, an insulator, while the P and N type material act as the plates of a capacitor. If the amount of reverse bias across the P and N type material is increased the depletion layer becomes wider giving the effect that the plates of the capacitor have been moved further apart so the value of capacitance reduces. If the amount of reverse bias across the junction is reduced the effect is the same as moving the plates of a capacitor closer together so the value of capacitance increases. It is possible using this device to create a variable capacitance by changing a voltage. This method is commonly employed in television and radio tuners to change the operating frequency using a dc voltage rather than using a mechanically variable capacitor. Its circuit symbol is shown in Figure 10.11.

Figure 10.11
Varicap diode

The range of capacitance varies from device to device but the typical characteristics are shown in Figure 10.12.

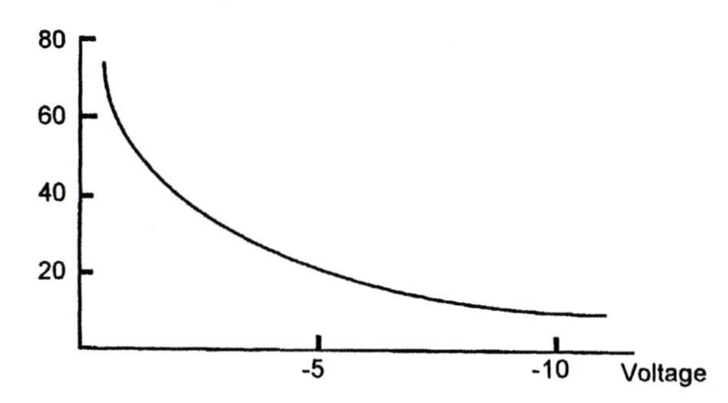

Figure 10.12
Varicap diode characteristic

Light emitting diodes – LEDs

This type of diode is used as a general indicator and in many devices has replaced the small incandescent lamp. The advantages of this device when compared to the lamp are in their lower operating voltages and currents and that they do not employ a wire filament. They are available in a wide range of packages, from those containing only a single diode to those containing a number of diodes. Figure 10.13(a) shows the circuit symbol of a single diode and Figure 10.13(b) a diagram of its encapsulation which helps to identify its electrodes. LEDs are available in different colours: red, green, yellow, white and blue. To reduce the current flowing through the LED a resistor is connected in series with it. The value of this resistor can be determined by applying the formula:

$$\text{Resistance} = \frac{V - V_f}{I}$$

where V is the supply voltage, V_f the diodes forward voltage drop and I is the maximum forward current. Both V_f and I will be quoted in the diodes characteristics. Consider the example shown in Figure 10.14 where the diode's maximum forward current is 30 mA and its forward voltage drop 2.1 V. This results in a series resistor value of 330 Ω.

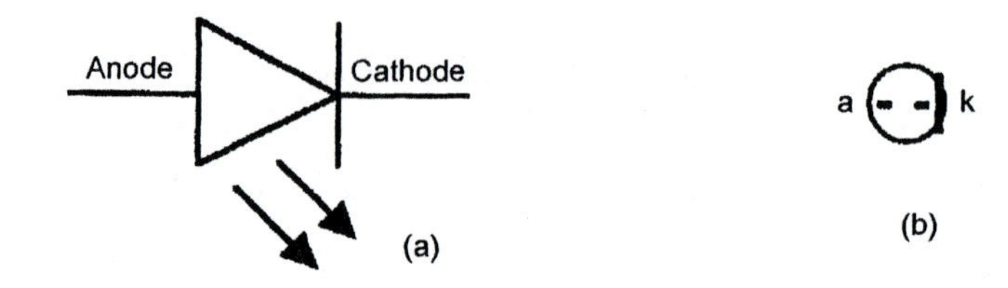

Figure 10.13
(a) Light emitting diode.
(b) Encapsulation

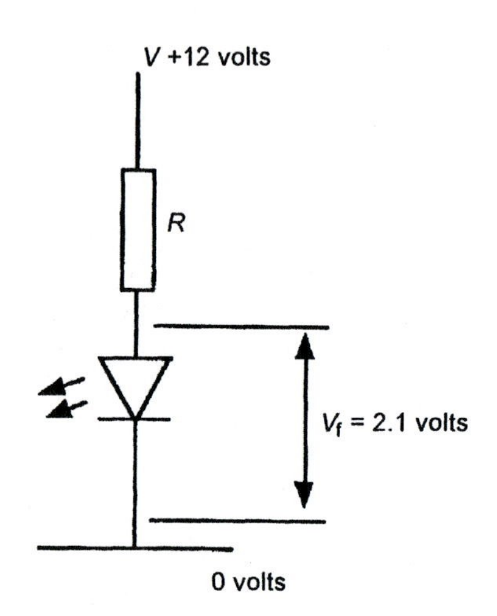

Figure 10.14
LED and series resistor

$$R = \frac{V - V_f}{I} = \frac{12 - 2.1}{30 \text{ mA}} = \frac{9.9}{30 \text{ mA}} = 330\ \Omega$$

It is also worth noting that yellow and green devices generally produce less light output for a given forward current than the standard red LED. This can be compensated for by reducing the series resistor of the yellow/green diodes or increasing the value of the series resistor of the red diode by 10–15%.

Bipolar transistors

The P and N type material manufactured for the production of the P–N junction diode can also be used to form another set of semiconductor devices known as transistors. The diode made use of a single P–N junction but the transistor is formed using two junctions as shown in the diagrams of Figure 10.15(a) and (b).

This type of transistor is termed bipolar because it makes use of two junctions. In either of the examples shown in Figure 10.15(a) and (b) the material sandwiched in the middle is named the *base* electrode whilst the outer blocks are named the *collector* and the *emitter.* It appears from the information given so far that the outer blocks can be either the emitter or the collector but this is not so because of the methods used to manufacture the device. The simple block structure shown in Figure 10.15(a) and (b) is used only as an aid to understanding how a transistor is constructed and how it functions. Since the transistor can be constructed using the blocks NPN or PNP it is necessary to have two different circuit symbols. These are shown in Figure 10.16(a) and (b).

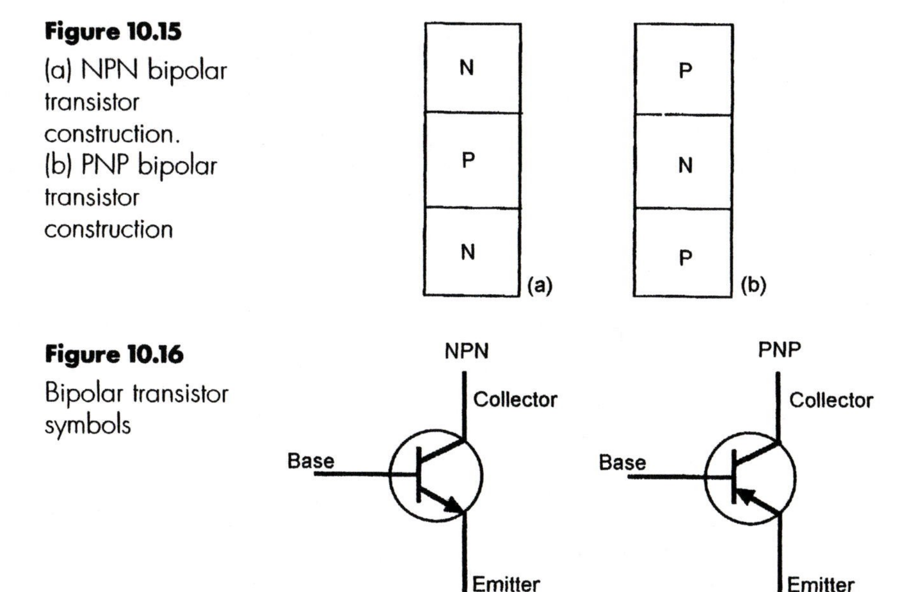

Figure 10.15
(a) NPN bipolar transistor construction.
(b) PNP bipolar transistor construction

Figure 10.16
Bipolar transistor symbols

The arrowhead is used to identify the emitter electrode and also the type of construction used to manufacture the transistor. When the arrow on the emitter is pointing outwards the transistor is an NPN construction and when the arrow is pointing inwards the transistor is a PNP construction.

To enable a transistor to operate it is necessary to provide it with the correct dc bias voltages. The correct bias conditions for either a PNP or an NPN transistor are that the base emitter junction must be forward biased and the base collector junction must be reverse biased. This is illustrated in Figure 10.17 and uses batteries to help simplify matters, but in practice the correct bias conditions will be provided by other means which will be explained later.

Battery 1 provides the base emitter junction with forward bias. The P type base material is connected to the positive terminal of the battery while the N type emitter material is connected to the negative terminal of the battery. Battery 2 has its positive terminal connected to the N type collector region and its negative terminal connected to the P type base material. The base emitter junction is forward biased and the base collector junction is reverse biased. Using the information given previously about semiconductors it reveals that current should be flowing through the forward biased junction, the base emitter junction, and no current should be flowing through the reverse biased base collector junction. However, the physics of the transistor is more complex and not essential for those just beginning to learn about electronics, or necessary to appreciate how a transistor can be used. It can be suggested that the transistor acts as a variable resistor and its name comes from a shortened version of the phrase *transfer resistor*. This is also helpful in understanding some of its actions. A more detailed explanation of the bipolar transistor as an amplifier is given in Chapter 12. Another type of transistor called a *field effect transistor* is also available.

Figure 10.17
Biasing for an NPN transistor

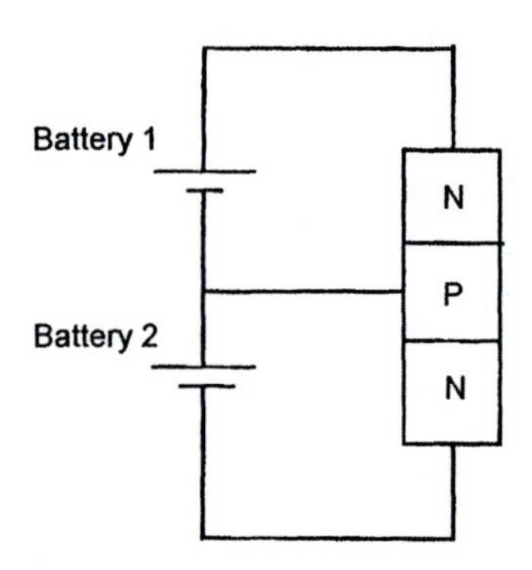

Field effect transistors

The junction gate field effect transistor JFET

Pure semiconductor material such as silicon or germanium can become relatively good conductors if they are doped with N type or P type impurities. The depletion layer formed between the N type and P type behaves as an insulator and controls the amount of current that flows through the device.

Examining the schematic diagram of a JFET shown in Figure 10.18 shows that the conducting channel which passes between the two electrodes, known as the source and the drain, can be varied in width by changing the thickness of the depletion layer connected to the electrode called the gate. The gate effectively acts as a clamp would around a hose pipe. If the clamp is tightened the hole through which water can pass is smaller so the amount of water

Figure 10.18
Schematic diagram of an N channel JFET

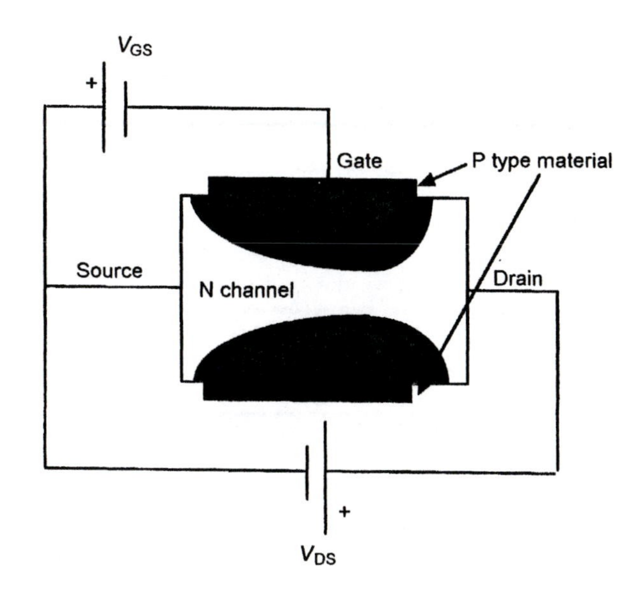

flowing is reduced. The flow of electrons is controlled by applying a negative potential to the gate (in this N channel device) to form a depletion layer in the device. The depletion layer is not uniform in width since it is widest near to the drain electrode where the potential difference is greatest. As a result of the formation of this depletion layer the electrons passing from the source to the drain have to pass through this conducting region whose width is controlled like the clamp around the hose pipe. Changing the voltage at the gate with respect to the source potential achieves this. If the gate–source potential is increased the current becomes smaller because the cross-sectional area of the conductor has reduced. This potential can be increased to a point where current ceases to flow and this is called the *pinch off voltage*, V_p. The conducting channel can be formed from either P type or N type material and to distinguish between them different circuit symbols shown in Figure 10.19 are used.

Figure 10.19
Circuit symbols for JFETs

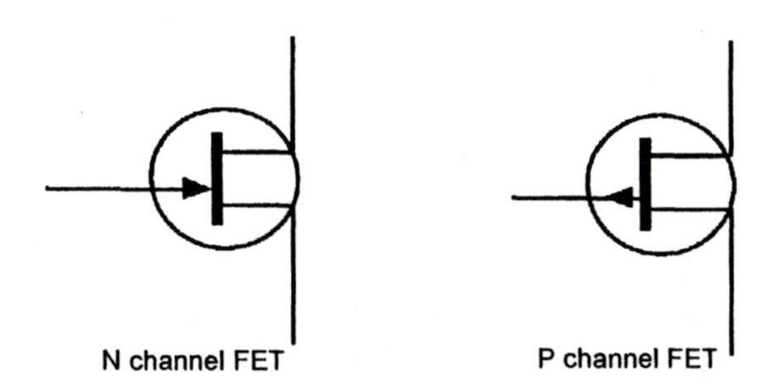

MOSFETS

In addition to the JFET there is another family of devices known as MOSFETs, *metal oxide field effect transistors.* In this type of device the gate region is separated from the conducting channel by a thin insulating surface which gives rise to the title *insulated-gate FET.* The depletion layer is created in this device by the charge created between the differing voltages on the conducting channel and the gate region. An N channel IGFET, as this type of device is known, is shown in Figure 10.20.

Figure 10.20
N channel IGFET construction

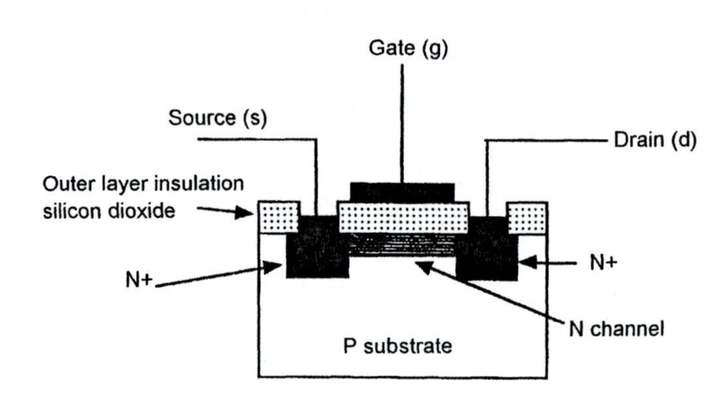

There are two commonly available types of MOSFETs. Those which operate in *depletion mode* and those which operate in *enhancement* mode. In the depletion mode the voltage applied to the gate reduces the amount of current flowing through the conducting channel. It could therefore be said that bias is needed to reduce the amount of current flowing, whilst in the enhancement mode forward bias is needed at the gate electrode to make current flow. The circuit symbol for each of the devices is shown in Figure 10.21.

Breakdown of a JFET device will occur if a sufficiently high voltage is applied between the drain and gate. The device will be destroyed by the heat that is developed within the transistor. Breakdown of a MOSFET will occur if the voltage applied to the gate is sufficiently large to cause a breakdown of the thin insulating layer between the gate and the conducting channel. This voltage may be the result of a static discharge caused by the type of clothing worn by an individual being placed across the electrodes. Voltages in excess of a thousand volts can easily be built up in this way. Protective packaging should always be used along with earthing wristbands when working on equipment using this type of transistor. It is important to ensure that your soldering iron is earthed so that no leakage can occur to damage the device.

Figure 10.21
Symbols and connections for various types of IGFET

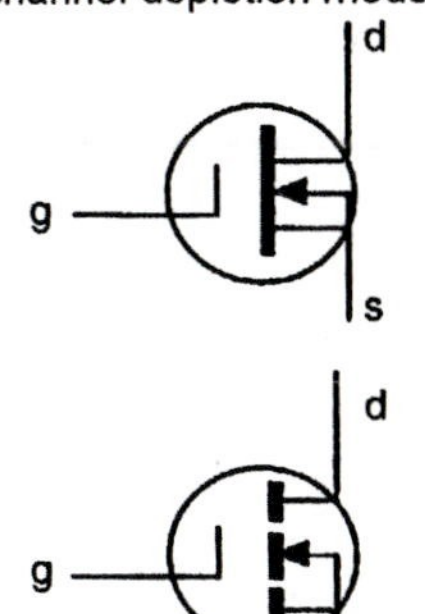

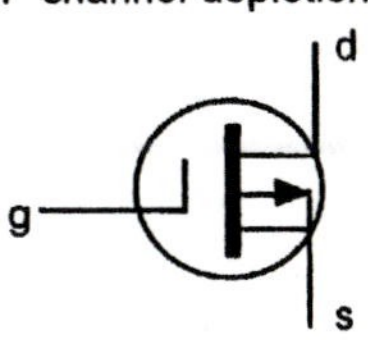

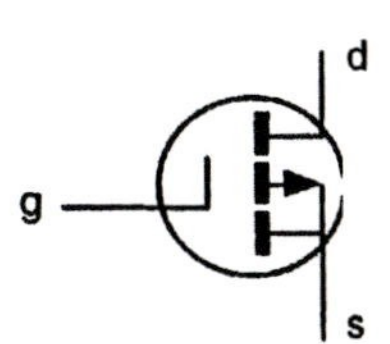

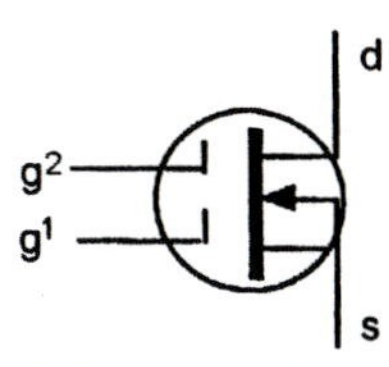

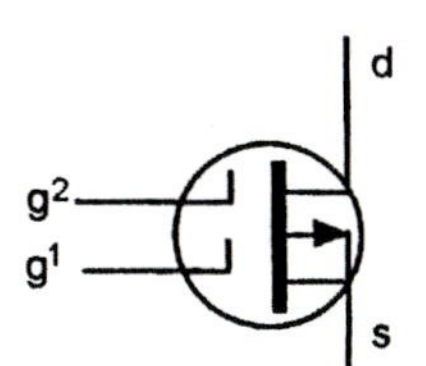

FET parameters

The gain produced by a field effect transistor is usually quoted by a figure stating its forward transfer conductance g_{fs} when connected as a common source amplifier. This mode is where the input signal is applied to the gate and the output current appears in the drain. The amount of gain being equal to:

$$g_{fs} = \frac{I_d}{V_{gs}}$$

where I_d is the change occurring in the drain current and V_{gs} the voltage causing the change in drain current.

When using FETs it is important to note some other parameters provided by suppliers.

$I_{D\ max}$ – The maximum drain current

$V_{DS\ max}$ – The maximum drain-source voltage

$V_{GS\ max}$ – The maximum gate-source voltage

$R_{DS(on)max}$ – The maximum value of resistance between drain and source when the transistor is conducting.

Bias

As with bipolar transistors it is necessary to apply the correct bias voltage. The amount and type of bias will be determined by the mode of operation but will necessitate the application of a gate to source bias typically between −2 V and +2 V. The circuit diagram of an FET used as an amplifier is given in Chapter 12.

Thyristors

These devices are also known as *silicon controlled rectifiers*. They are used as a switching rectifier particularly in ac power control circuits. They have the ability to switch large currents very quickly and of course they do not require moving contacts as do relays, etc. When switched off the thyristors resistance is high and there is very little leakage current, and when the device is conducting its resistance is very low so little power is lost even if a large current is flowing. Figure 10.22 shows the circuit symbol for a thyristor.

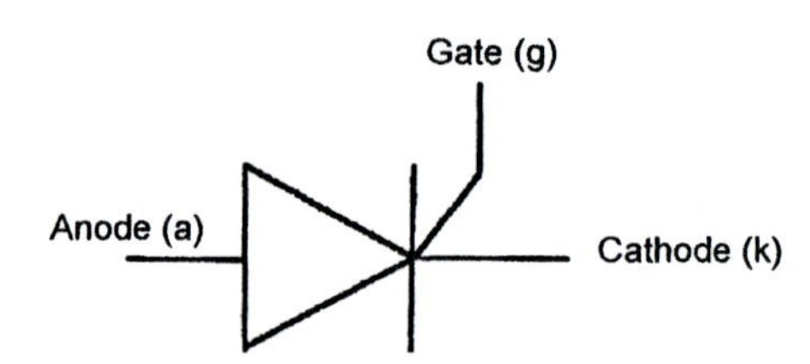

Figure 10.22 Circuit symbol for a thyristor

To make the device conduct its anode must be positive with respect to its cathode and a positive control pulse applied to the gate. When these conditions are met the thyristor turns on. It will remain conducting until the forward current falls to an amount less than the holding current usually when the anode voltage has fallen to a level almost equal to the cathode voltage. The simple circuit in Figure 10.23 shows how a thyristor is used to produce a dc output from an ac input.

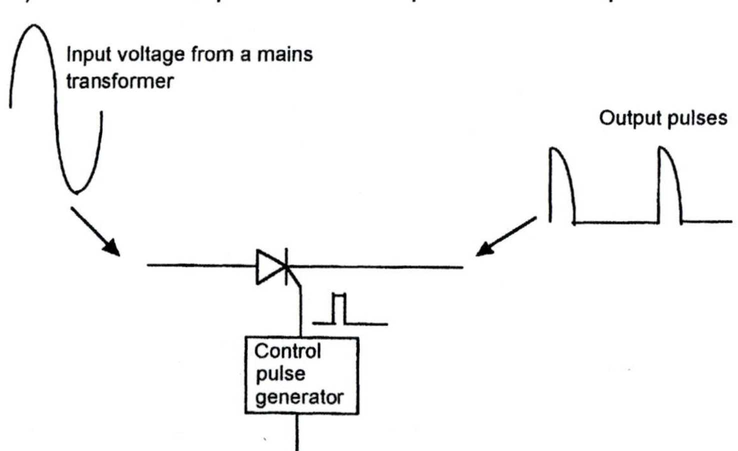

Figure 10.23 Thyristor rectifier circuit

The ac input is applied to the thyristor anode. The thyristor cannot conduct until a positive pulse is applied to the gate. The time during the positive half cycle of the input at which this pulse is applied is controlled by the pulse generator. If the pulse is applied early in the positive half cycle of input then the thyristor conducts for longer than when the control pulse is applied later. The effect of this is to enable the amplitude of the output voltage to be controlled. This gives the thyristor an advantage over a conventional rectifier whose output is totally dependent on the input voltage.

The pulses of output voltage must be smoothed if the output desired is a dc voltage. Thyristors can as previously stated be used for switching large or small currents and just as with other devices the physical size and construction of the device are determined by the current that it is required to pass. A thyristor used in a circuit where a large current is flowing is generally made from a metal package and could be mounted to a heat sink, but where only small currents are passed the thyristor will be a small plastic package and have the same appearance as a small transistor.

11

Electrical components

Most of the common components used in electronic and electrical equipment, i.e. resistors, capacitors, inductors, etc., have been considered in their own dedicated chapters. However, there are countless other components and devices employed in the electronics industry, and it would be very difficult to look at every one of these in a single book. In this chapter we will consider just a few more common and essential components.

Fuses

The electrical and electronics industries employ many types of fuse and fusible devices to protect people from electrical shock, to reduce the risk of electrical fires, and to protect electrical and electronic circuits from damage. In Chapter 1 we saw how a fuse can prevent a person from experiencing an electric shock by rupturing and breaking the circuit in the event of a fault. We shall now look at the different types of fuses that are available, and consider the circuit protection qualities.

The conventional cartridge fuse comes in a number of sizes, construction types, and current ratings. The small glass 20 mm version is used extensively in electronic equipment; however, these are only available at low current ratings (up to around 15 A). At the other end of the range are the large cartridge types capable of handling hundreds of amps.

The fuse wire is mounted into a cartridge not simply for the convenience of fitting. When a current value greater than the fuse rating flows through the wire, the temperature of the wire rises to a few thousand degrees, causing it to evaporate. This action is very rapid and the wire can explode, spraying molten metal over a considerable distance. The glass or ceramic cartridge is intended to contain the explosion, preventing both injury and damage to equipment.

As well as having a current rating, a cartridge fuse also has a *voltage rating*. This refers to the maximum working voltage that can be applied to the fuse without danger of flashover when the fuse is caused to rupture. Flashover must be avoided for two reasons: the cartridge may explode causing injury, and circuits or components, especially silicon devices, may be damaged. A replacement fuse should have a voltage rating at least equal to the applied voltage. A 125 V fuse must not be inserted into a 230 V circuit. The most common voltage ratings are 125 V and 250 V, although 32 V ratings are commonplace in the automotive industry.

The *speed of the fuse* refers to the time it takes for the fuse to rupture. Heating and rupture is not instant, and although one could be forgiven for thinking that it is essential for the fuse to rupture as quickly as possible, this is not always advantageous. Some equipment draws a high current, in excess of the normal operating current, for a brief period at switch-on. In such cases the fuse must be able to withstand the surge, but must rupture in the event of a fault condition that would result in a sustained excess current. These fuses are known as *time delay or anti-surge fuses*.

Fuses are designated according to their operating speed. FF indicates a very fast-acting fuse, commonly employed to protect semiconductor devices. F denotes a normal acting fuse with no delay period. M is a medium time lag fuse that will withstand a very brief surge current. T (known also as 'slo blo') is the most common type employed in domestic electronic equipment. It offers sufficient delay to permit the charging of smoothing capacitors and the like, without compromising safety. TT indicates a very slow rupture time and will withstand a considerably high surge current. The delay designation is generally stamped into the metal end cap along with the current and voltage ratings.

Fuses may be colour coded. For example, the cartridge fuses employed in 13 A plugs are coded red for 3 A, black for 5 A, and brown for 13 A. It is sometimes confusing that the 2 A plug fuse is also coloured black.

Twenty millimetre glass cartridge fuses may have coloured stripes painted on to indicate their value. This code works in a similar way to the resistor colour code considered in Chapter 6, but in the case of a fuse the first two stripes indicate a value in milliamps, and the third band is the multiplier. For example, a fuse coded *red–black–red* would have a value of 20 mA × 100 = 2000 mA, or 2 A. If a fourth band is included, it would indicate the speed of the fuse; brown is FF, red is F, yellow is M, and blue is T.

When fault finding on an item of equipment that is completely dead, the main fuse should be one of the first items to be tested. Testing a fuse is very simple as the device should normally read almost a short circuit, that is, 0 Ω. To test a fuse, a multimeter switched to the lowest ohms range is used. Connect the fuse across the test leads and check that the reading is almost 0 Ω. For further instructions on using a multimeter for testing devices, see Chapter 15.

A fuse is normally either good or bad, i.e. short circuit or open circuit. If a test indicates that a fuse has resistance, then it will probably require replacing, although some fuses below 300 mA can have a resistance of a few ohms.

If the fuse is a clear glass type, then a visual inspection can be informative. If the wire appears to have broken in two, then there is a high probability that the fuse has failed through age. Basic checks should still be carried out on the equipment for short circuits, but it is often found that a replacement fuse will restore normal operation.

If the inside of a glass fuse is black, then it will have ruptured because of excessive current, and the equipment must be tested thoroughly for defective components before replacing the fuse and switching on.

Other fusible devices

Although the traditional wire fuse is an essential feature of all electronic equipment, there are a number of other types of protection device available whose properties offer advantages over the wired fuse in certain circumstances.

The *fusible resistor* functions well where current surge limiting is required in addition to a fuse. These devices look very similar to their carbon resistor cousins; however, they are designed to quickly go open circuit if their maximum power dissipation level is exceeded. A normal carbon resistor would take much longer to burn out, and there is a real risk of a fire during this period.

A typical application of a fusible resistor is shown in Figure 11.1. The 4.7 Ω resistor R_1 is being used as a surge limiter to restrict the initial charging current flowing into capacitor C_1 at switch-on. In the event of C_1 becoming short circuit, or a fault developing in the 'main circuit' that would cause an excessive current flow, R_1 will heat up and burn out, thus acting as a protective fuse.

Figure 11.1

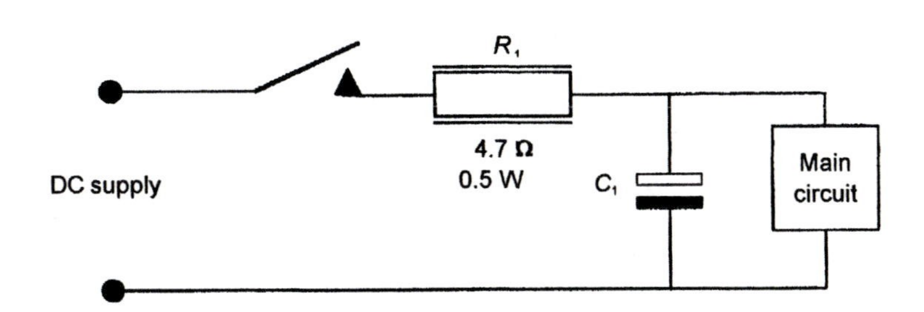

It is important to note that when replacing a defective fusible resistor, a component with the identical value and power rating must be used. A carbon resistor must never be used as a replacement as this would be the same as fitting a 6" nail in place of a glass wired fuse!

Another type of protective device is the *thermal fuse*. This comes in different guises. Two types are illustrated in Figure 11.2.

The ceramic wirewound fusible resistor operates by heating a spring-loaded contact that is held closed by high melting point solder. Under no-fault conditions the device functions as a normal wirewound power resistor. However, in the event of a fault condition that results in an excessive current flow, the temperature of the resistor rises rapidly, melting the solder and allowing the contacts to spring apart. The contacts are connected in series with the resistor, so when they break, the circuit is isolated. One drawback with these devices is that there may be a considerable time delay before the 'fuse' blows.

The pellet type fuse shown in Figure 11.2(b) is made of a metal alloy that has a specific melting temperature. A device such as this may be found in electrical appliances that operate at high temperatures, e.g. storage heaters. Typical ratings are between 1 A and 15 A, with operating temperatures between 60°C and 150°C.

A smaller version of the pellet fuse is often found built into mains transformers. Should the transformer overheat the pellet will melt, breaking the primary circuit. However, in many cases it is not possible to replace the fuse as it is an integral part of the transformer construction. This usually means that a repair is more expensive because the entire transformer must be replaced.

A more recent addition to the family of fusible protection devices is the *integrated circuit protector* (or *ICP*). These tiny fast acting solid state devices are designed to fit onto a printed circuit board, and offer similar protection to an F rated cartridge fuse. They are not designed to operate above 50 V dc.

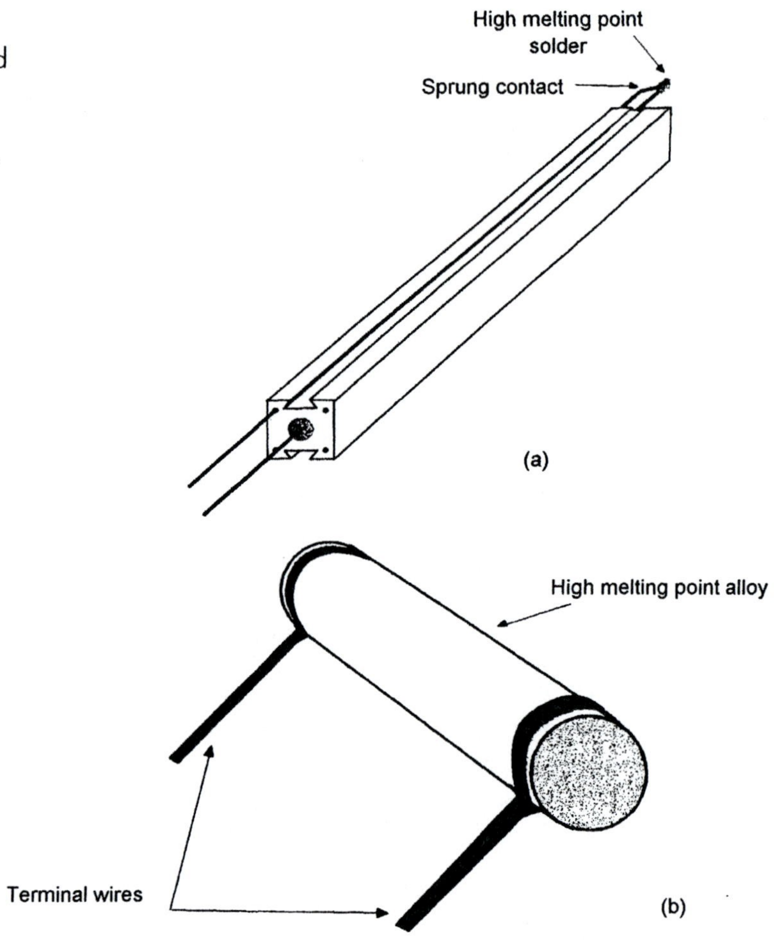

Figure 11.2
(a) Wirewound resistor incorporating a thermal fuse. (b) Pellet type thermal fuse

ICPs frequently use the same encapsulation as some transistors, and at first glance may be identified as such. However, closer scrutiny reveals only two legs. A typical encapsulation is shown in Figure 11.3.

Figure 11.3
Typical integrated circuit protector

The current rating of an ICP is determined by multiplying the last two digits of the type number by 40 mA. For example, a device labelled ICP-F10 would be a 400 mA fuse with an F rating. An ICP is tested in the same way as any other fuse, using a multimeter.

The British Standard symbols for different fuses and fusible devices are given in Figure 11.4.

Figure 11.4
Fusible resistor symbols

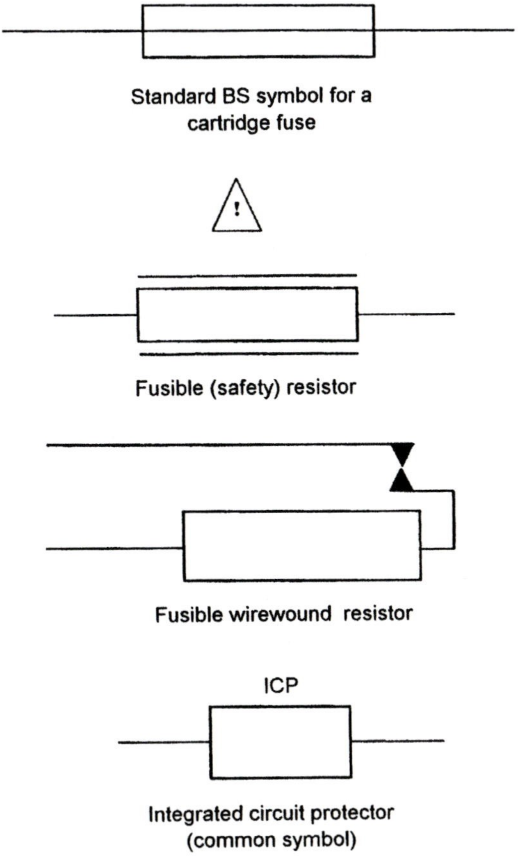

Standard BS symbol for a cartridge fuse

Fusible (safety) resistor

Fusible wirewound resistor

ICP

Integrated circuit protector (common symbol)

Filament lamps

The principle of operation is very simple. A wire filament, usually tungsten, is made to have a resistance of a few ohms so that when a current is passed through it, power is dissipated (remember that Power = $V \times I$). When connected into a circuit, the filament behaves as a resistance, burning white hot and hence emitting light. If oxygen were present the filament would rapidly burn out, so it is suspended in a glass bulb containing an inert gas. The gas also helps to even the pressure on the glass lamp, and it contributes to the light output when heated.

The filament has a positive temperature coefficient of resistance which means that its resistance, when tested on a meter, would read much higher when the lamp is warm than when measured cold. The low cold resistance means that there is a high surge current at switch-on, and this current can be as high as ten times greater than the normal operating current.

There is a vast range of filament lamps available, offering a wide choice of physical size, light output, colour temperature (in simple terms, the colour of white light output, i.e. yellow, blue), power consumption, reliability, and of course cost.

Filament lamps can have a long life; however, this can be drastically shortened by such things as extreme ambient temperature, excessive operating voltage, and mechanical vibration, especially when hot. The type of supply voltage also has an effect on lamp life. Lamps generally have a longer life when operated from an ac source. When a dc source is used, an effect known as tungsten migration occurs in the filament which can reduce its life.

The optimum ambient operating temperature is typically quoted as being between 20°C and 25°C, although many lamps can function with reasonable reliability at temperatures outside of this range.

Mechanical vibration is often unavoidable, for example where the lamp is used in a motor vehicle or a hand torch, and it must be accepted that these lamps may not endure as long as those used in static equipment.

The voltage at which a lamp is operated will be determined by the engineers designing the equipment in which the lamp is to be used. They can choose to operate the lamp at its specified voltage which will in theory offer optimum light output and reliability. However, in practice this is not always the case because the power supply rail may be prone to variations, and the other factors, vibration and ambient temperature, must be taken into account. A 5% increase in operating voltage can reduce the life of the lamp by about 50%, whereas a 5% reduction in operating voltage can significantly increase the life expectancy. So it can be seen that the design engineer can compensate for adverse operating conditions by reducing the operating voltage of a lamp, at the expense of some light output. The effects on lamp life and light output for different supply voltages are illustrated in Figure 11.5.

Filament lamps are generally rated in watts; however, the rating of some smaller lamps is given in amps. Where the supply voltage is higher than the maximum rating of the lamp, the current can be limited to the correct value by employing a series resistor. Both the value, and the power rating of the resistor must be calculated.

Figure 11.5
Effect of operating voltage on lamp performance

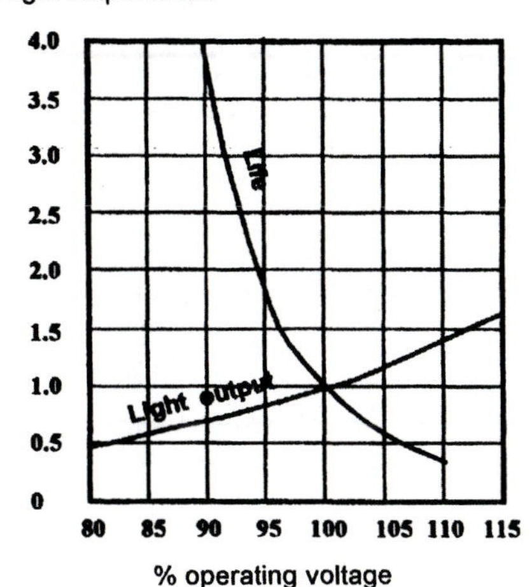

Example

The circuit in Figure 11.6 illustrates how a 12 V lamp is connected to a 20 V supply using a series current limiting resistor. Calculate the optimum resistor value and power rating for reliable lamp performance.

Solution

p.d. across $R_1 = 20\ \text{V} - 12\ \text{V} = 8\ \text{V}$

$$\therefore R_1 = \frac{8}{0.1} = 80\ \Omega$$

Power $= 8 \times 0.1 = 0.8$ W

Using common preferred values, a 91 Ω 1 W resistor would be a good choice.

Figure 11.6

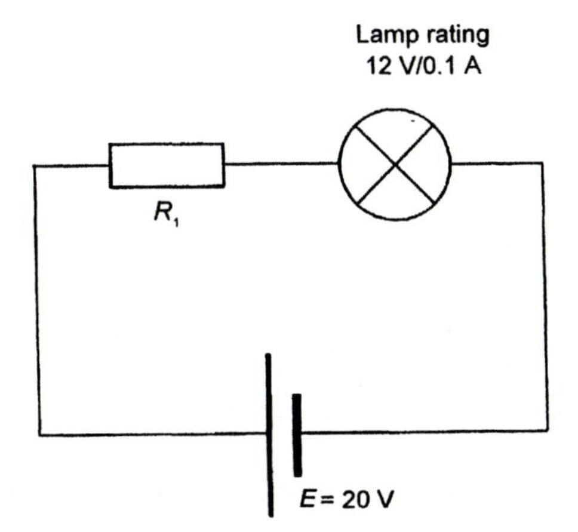

Where illumination is required in extreme temperature conditions, for example in a refrigerator, oven, or microwave oven, special types of lamp are used. These use a glass envelope that can withstand not only extreme temperatures but also sudden temperature changes.

Filament lamps are employed in electronic equipment for panel illumination, indication, or decoration. Miniature lamps are available in a wide variety of voltages ranging from very low, in the order of 3 V, to high, around 130 V. The range includes different physical construction and fitting. Some common types are: Miniature Edison Screw (MES), Lilliput Edison Screw (LES), Miniature Bayonet Cap (MBC), Flanged, and Wedge base. These are illustrated in Figure 11.7.

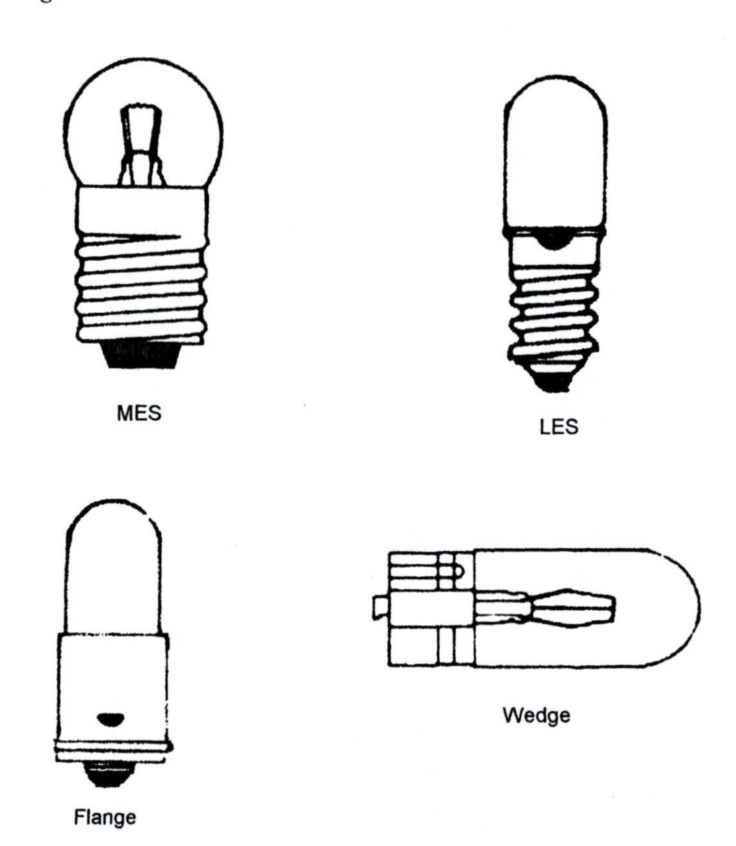

Figure 11.7 A range of common filament indicator lamps

Apart from connecting it to a known good supply, a suspect lamp can be tested either by visual inspection of the filament, or by connection to an ohmmeter.

If a visual inspection reveals that the filament is broken, or the glass has burn marks on the inside, then the lamp must be replaced. When testing a filament lamp, an ohmmeter should give a reading of a few hundred, or perhaps just over a thousand, ohms for a good lamp. The expected resistance reading can be calculated using Ohm's law if the rating of the lamp is known. If the lamp is defective it will read either open circuit, or very high resistance.

Switches

There are countless types of mechanical switch, e.g. sliding, rocker, rotary, single pole, double pole, multipole, low current, high current, low voltage, high voltage, and so on. Switches are a common cause of trouble in electronic equipment because they can be physically broken, and the contacts can burn out or become corroded. This can lead to all types of faults in the equipment, including intermittent ones, and it is important that a service engineer can interpret service information in order to identify specific switch contacts so that resistance checks can be performed.

When a switch is operated, contacts move in order to make and/or break the contact connections. Many switches are multipole, which means that they contain more than one set of contacts. Such switches are defined as: Single Pole Single Throw (SPST), Single Pole Double Throw (SPDT), Double Pole Single Throw (DPST), and Double Pole Double Throw (DPDT). The contact arrangement for each of these is shown in Figure 11.8, along with some other common switch types.

Perhaps the simplest switch is the toggle type, where a sprung paddle is operated to throw the contacts from one position to the other. A typical example is shown in Figure 11.9, which illustrates a single pole double throw (changeover) type, although these are available in the four types illustrated in Figure 11.8.

Surface mounted devices (SMDs)

Since the birth of the electronics industry designers have been striving to reduce the size of equipment. Although a lot of the pressure for size reduction came from the military and space technology quarters, even domestic consumers were applying a pressure by indicating that, given the choice, they would purchase the smaller of two identical items.

The size of a piece of electronic equipment is governed very much by the size of the components used in its manufacture, and over the years all of the major components have been reduced in their physical size. However, there is a limit as to how much the size of a component can be reduced before traditional assembly methods become impractical. For example, if a resistor is only 1.5 mm in length, it is not practical to drill two adjacent holes of 1 mm diameter to accommodate the leads. And when this problem is multiplied a thousand times on one printed circuit board (PCB), it follows that you will be left with more or less one large hole!

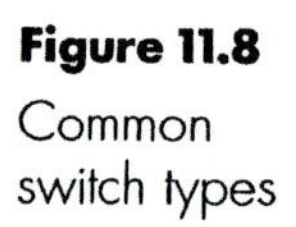

Figure 11.8
Common switch types

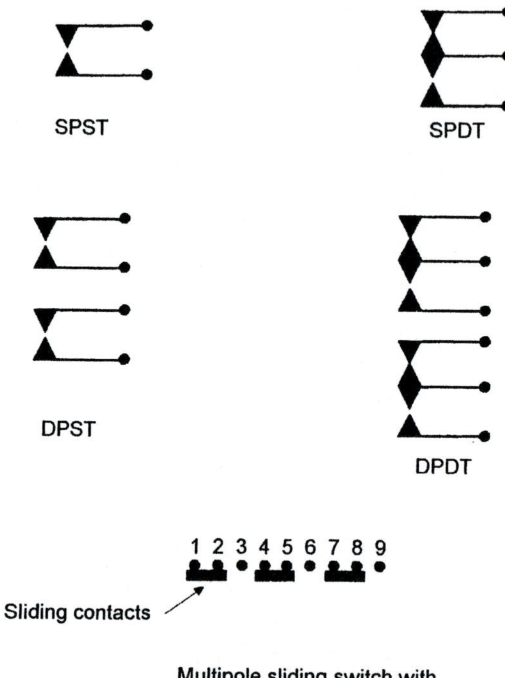

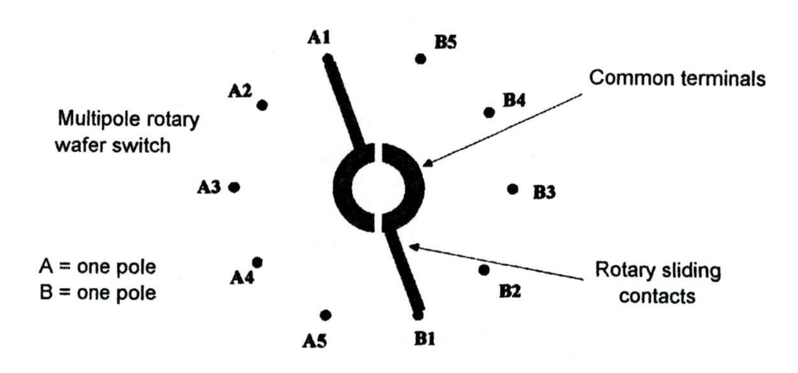

As well as reducing production size, manufacturers were also looking into ways of reducing production costs. During the 1960s and 1970s electronic equipment assembly was very labour intensive, requiring hundreds of people on long assembly lines to produce just one model of equipment. Production costs would be greatly reduced if these people could be replaced by a few machines (this is not the place to discuss the social or political issues involved!). One problem was the component size; in general they were too large to be fitted rapidly by machine.

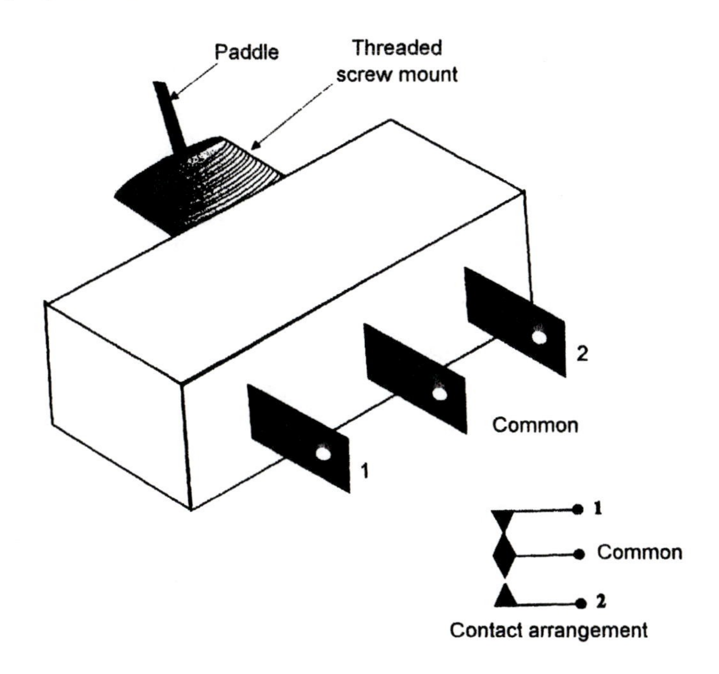

Figure 11.9 A single pole changeover toggle switch

The solution to the problems relating to component mounting and production costs was found in the surface mount technique, where a component is simply stuck onto the PCB and then flow soldered. During the 1980s a new breed of component was developed. Resistors and capacitors of all values, and all types of semiconductors including silicon chips, were reduced to just a fraction of their size. These components are small enough to be fitted into bandoleers which in turn are mounted into automatic insertion machines. The majority of components can be fixed onto a single large PCB in a matter of minutes, and furthermore, the machine is capable of testing each device before it is mounted, rejecting it if it is found to be defective or out of tolerance. Because the board does not have to be drilled to accommodate the components, both sides of the PCB can be used, thus allowing for further size reduction.

From a servicing point of view SMD brings new challenges. Attacking an SMD IC with a traditional 25 watt large soldering iron will usually result in irreparable damage to the PCB. Locating a resistor that is not much larger than a piece of household dust is very difficult unless there is a lot of illumination, and even then a magnifier is recommended. The modern electronics servicing workshop requires special de-soldering tools and equipment to handle SMD technology boards, but armed with such equipment the modern service engineer is still able to fault find and repair to component level.

12

Electronic circuit applications

The transistor as an amplifier

The bipolar transistor described in Chapter 10 can be used in a number of ways, and acting as an amplifier is one of its most important functions. If a transistor is connected in the experimental circuit shown in Figure 12.1 some important facts are revealed. This circuit shows the transistor, an NPN type, correctly biased. The forward bias of the base emitter junction is achieved by applying a positive voltage through ammeter A_3 and variable resistor VR_1. VR_1 is used to set the amount of current flowing in the base emitter junction which is termed the base current I_b. The output current is measured by ammeter A_1 which measures the current flowing in the collector through resistor R_2 which is included to keep the current within the limits of the transistor's capabilities and prevents it being damaged. This current is called the collector current and given the symbol I_c. Since all the current flowing in the transistor has to pass through the emitter, ammeter A_2 can be used to measure the emitter current. When the supply is connected and current flows the currents measured on the three ammeters reveal that the emitter current I_e is equal to the base current I_b plus the collector current I_c. It also reveals that current is flowing through the base collector junction which is reverse biased. This can be explained at this time by the simple suggestion that as the electrons are moving from the emitter into the base region, attracted by the positive voltage on the base which is only a thin layer, the electrons burst through it and having done so are attracted to the collector because it is more positive than the base and therefore more attractive to a negatively charged electron.

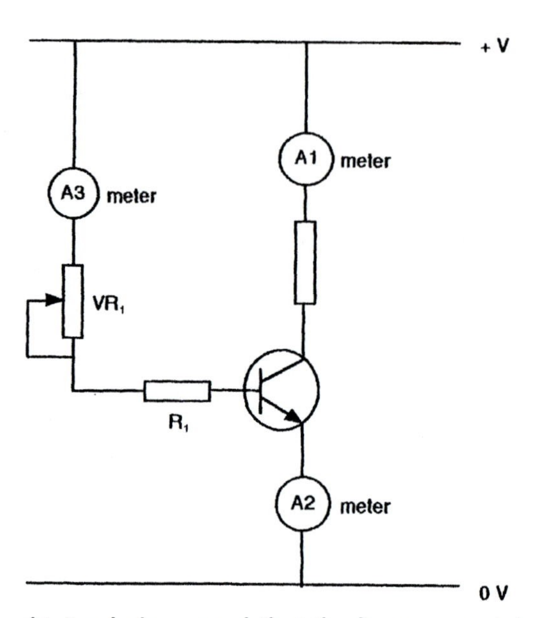

Figure 12.1 Simple transistor amplifier

The readings obtained also reveal that the base current I_b is much smaller than the collector current I_c. As a rough guide it can be suggested that the total current, the emitter current I_e represents 100% whilst the collector current I_c 98% and the base current I_b 2%. Using these figures the conclusion that can be reached is that a base current of 2 mA results in a collector current of 98 mA. If VR_1 is adjusted to give a reading of 1 mA on meter A_3 the reading now obtained on meter A_1 will be 49 mA. This is based on the statement made earlier that the base current had a value of 2% of the total current and the collector current 98% of the total current. If VR_1 is readjusted so that the base current reading on meter A_3 is 3 mA the reading now obtained on meter A_1 will be 147 mA. The base current has changed by 2 mA but the collector current has changed by 98 mA. A small change in current at the base produces a much larger change in current in the collector, so the device has a *current gain*. This makes the transistor an extremely useful device and is used as an amplifier because a small changing current at its input can produce a much larger changing current at its output. This gain is referred to as the large signal current gain and is stated by the figure quoted as the transistors H_{FE}. This is a simple calculation and for the example given would be:

$$H_{FE} = \frac{\text{change in collector current } I_c}{\text{change in base current } I_b} = \frac{147\text{ mA} - 49\text{ mA}}{3\text{ mA} - 1\text{ mA}}$$

$$= \frac{98\text{ mA}}{2\text{ mA}} = 49$$

This is a simple example used for illustration. Transistors are manufactured with current gains that can be much higher or lower than the example shown. Figure 12.2 shows the simplest amplifier circuit possible. Normally the circuit needs to contain a few more components for better performance. The resistor R_1 provides the forward bias for the base emitter junction to turn the transistor on while R_2 provides the bias voltage on the collector. Since this voltage is higher than the base voltage the transistor has the correct bias condition as stated previously, reverse bias. Resistor R_2 is also known as the collector load resistor.

Figure 12.2
Common emitter amplifier

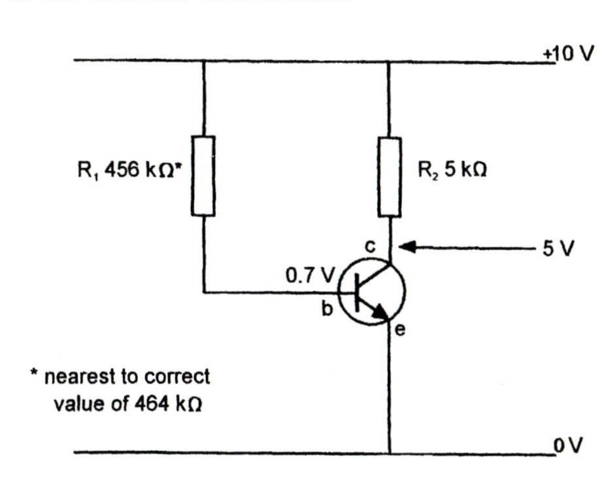

The value of resistor chosen provides a base current of approximately 20 μA and a collector current of 1 mA. These figures have been chosen simply to help gain some understanding. The base voltage will be 0.7 V the forward bias voltage of a P–N junction and the collector voltage will be 5 V if a 10 V supply is used. If a small alternating signal with a peak to peak amplitude of perhaps 50 millivolts is applied to the base electrode along with the dc bias the resulting base current will rise and fall in sympathy with the signal. If the base current is made to change, then as stated earlier it will cause the collector current to change by an amount determined by the current gain of the transistor. In the example of Figure 12.2 the H_{FE} is 50 so if the alternating signal changes the base current by 5 μA in the positive direction and 5 μA in the negative direction. The change in base current is 10 μA. With a current gain of 50 the change in collector current will be 500 μA so the maximum collector current will be 1.5 mA and the minimum collector current will be 0.5 mA, a change of 1 mA. The voltage on the collector can be determined using Ohm's law. When the collector current is at 1.5 mA the voltage dropped by R_2 will be 7.5 volts and when the collector current is 0.5 mA the voltage dropped by R_2 will be 2.5 volts. A 20 μA change in current at the base has resulted in a voltage change of 5 volts at the collector. To determine the voltage gain the output voltage is compared to the input voltage as follows:

$$\text{Voltage gain} = \frac{\text{Change in collector volts}}{\text{Change in base volts}} = \frac{5\ \text{V}}{50\ \text{mV}} = 100$$

An important point not to be missed is that the output voltage at the collector is in the opposite phase to the input voltage on the base. This is shown in Figure 12.3.

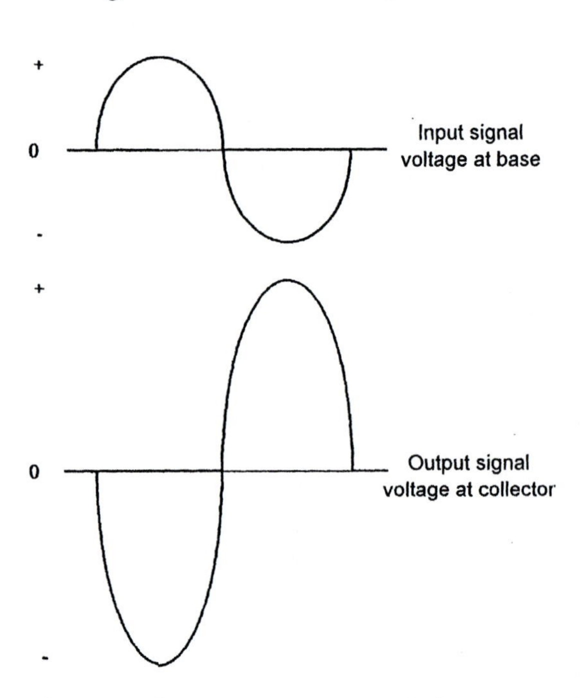

Figure 12.3
Input and output signals of a common emitter amplifier

The type of circuit shown in Figure 12.2 is known as a *common emitter* amplifier circuit. Although this circuit would function as an amplifier it would need some modification to improve its operation. A practical circuit is shown in Figure 12.4. The base bias needs to be a stable dc voltage since any changes would result in a change of current flowing through the transistor. A two resistor network, R_1 and R_2, connected as a potential divider is therefore used. A further problem is that of heat. As the transistor becomes warmer it passes more current which again produces a further increase in current and so on until the transistor destroys itself. This effect is called thermal runaway. To prevent thermal runaway a resistor is placed in series with the emitter. This resistor is known as the emitter stabilising resistor R_4. When the transistor starts to pass more current the voltage on the emitter developed across this resistor R_4 rises and becomes closer to the base voltage. The effect is to reduce the base bias potential and this holds the current steady. Unfortunately this resistor also has an affect on the ac signal current which is undesirable. It is necessary to remove the ac signal developed across the emitter stabilising resistor and to do this a capacitor is connected in parallel with the emitter stabilising resistor. This capacitor is known as the emitter bypass capacitor C_3.

Figure 12.4

Common emitter amplifier

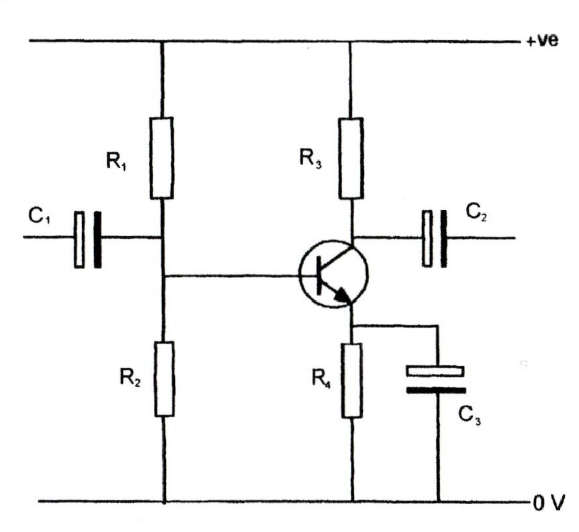

To prevent any external dc voltages affecting the levels set in the amplifier the ac signal is coupled via C_1 and C_2. These capacitors are called coupling capacitors. Their function being to couple the ac signal and block any dc voltage. Transistors can also be connected in different ways to carry out other tasks, but connecting them in alternative arrangements changes circuit parameters. Figure 12.5 shows the three methods of connection together for comparison.

The JFET as an amplifier

Figure 12.6 shows the circuit diagram of a simple JFET amplifier. It is the equivalent circuit to that shown for the bipolar common emitter amplifier and operates in a similar manner. The correct bias to the gate must be applied and this is with the gate less positive than the source. R_1 produces this condition. R_3 is the load resistor and determines the amplitude of the output signal for a given input signal and current. When the signal at the gate goes positive the depletion layer in the N channel is reduced so with a larger

Figure 12.5
Modes of connecting transistors as amplifiers

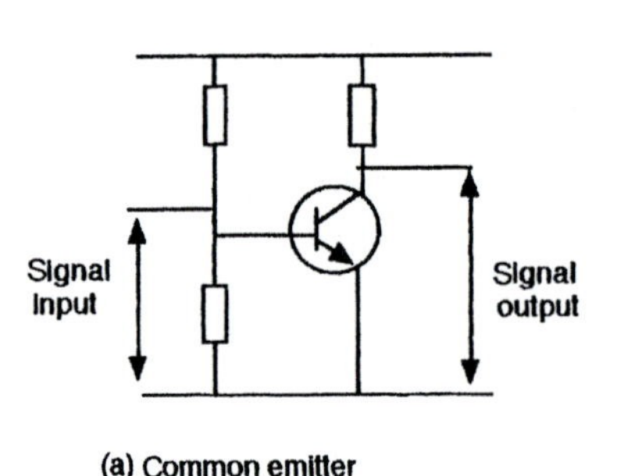

Voltage gain $= \frac{V_{out}}{V_{in}}$ = HIGH 100

Current gain $= \frac{I_c}{I_b}$ = HIGH 50–1000

Input resistance relatively low 1-5 kΩ

Output resistance relatively high 50 kΩ

(a) Common emitter

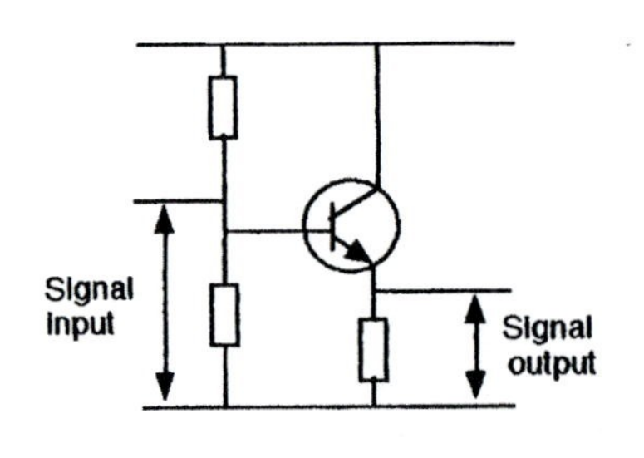

Voltage gain $= \frac{V_{out}}{V_{in}}$ = UNITY = 1

Current gain $= \frac{I_c}{I_b}$ = HIGH 50–1000

Input resistance HIGH several kΩ

Output resistance LOW few ohms

(b) Common collector or emitter follower

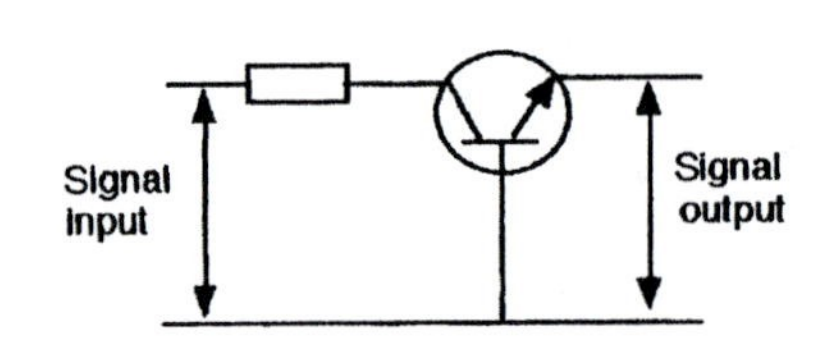

Voltage gain $= \frac{V_{out}}{V_{in}}$ = MEDIUM 50 to 100

Current gain $= \frac{I_c}{I_b}$ = UNITY = 1

Input resistance LOW 50 Ω

Output resistance HIGH 1 MΩ

Figure 12.6
FET amplifier

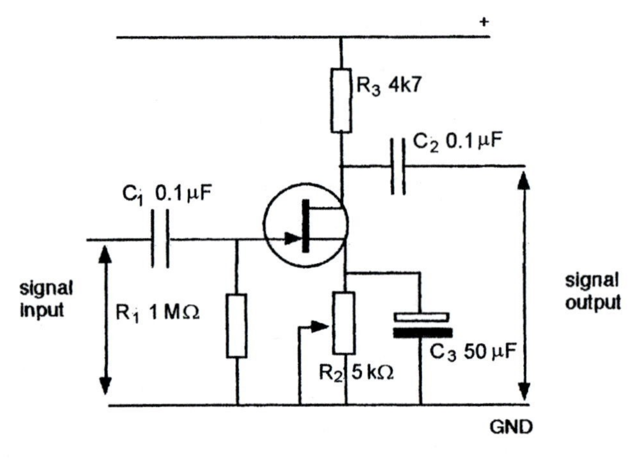

conducting channel more current flows resulting in a fall in voltage on the drain. When the signal at the gate goes negative the depletion layer increases so the conducting channel becomes narrower and less current flows. The voltage on the drain rises. Once again the signal on the drain is an inverted but larger signal than that at the gate. Just as in the common emitter amplifier the signal on the collector was a larger but inverted signal than that applied to the base. C_1 and C_2 are coupling capacitors. Their purpose is to allow the ac signal to pass but to block any dc voltage that may be present with the input signal which would change the dc bias conditions and possibly damage the FET. R_2 is a gain control since reducing its value allows more current to flow, whilst increasing its value reduces the gain because less current flows. C_3 prevents any ac signal being developed across R_2 and reducing the gain.

The transistor as a switch

In many applications transistors are used as switching devices where they behave as fully on or fully off devices. They are used in this mode of operation in squarewave and pulse oscillators shown in the following circuits.

The astable multivibrator

This is a free running oscillator which requires only a dc supply voltage to enable it to function. A typical circuit is shown in Figure 12.7.

Figure 12.7 Transistor astable multivibrator

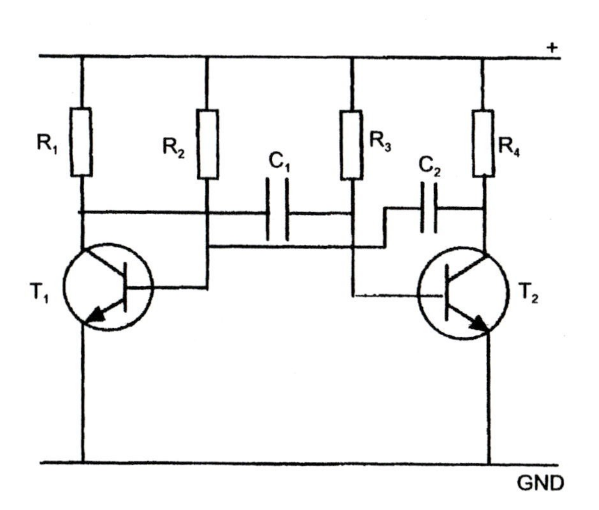

When first switched on one of the transistors will conduct before the other because of its characteristics, even though transistors of the same type are used. In the circuit shown assume that T_1 conducts first. When conducting the transistor is saturated, fully on, so its collector voltage is low. C_1 begins to charge through R_3 until the base of T_2 becomes positive at a level of approximately 0.6 V. At this level the voltage is high enough to turn on T_2 which now conducts. The resulting fall in its collector voltage charges the left-hand plate of C_2 negatively and this voltage turns off T_1. The collector voltage of T_1 rises. The transistors stay in this condition, T_1 off and T_2 on, as long as the negative charge remains on the left-hand plate of capacitor C_2. This negative charge leaks away through R_2 and as soon as the voltage at this point in the circuit reaches 0.6 V transistor T_1 conducts again. The falling collector voltage on T_1 results in capacitor C_1 becoming negatively charged on its right-hand plate, the plate connected to the base of T_2 switching off T_2. As long as C_1 holds the base of T_2 negative it cannot conduct and its collector voltage stays high. The charge on C_1 will leak away through R_2 and as soon as the base of T_2 becomes positive it will conduct again. The cycle repeats again. The rate of switching or oscillation is determined by the choice of components used for R_2, C_2, C_1, R_3. If R_2 is equal to R_3 and C_1 equal in value to C_2 the output waveform at the transistor's collector will be a symmetrical squarewave. The waveforms for such an oscillator are shown in Figure 12.8.

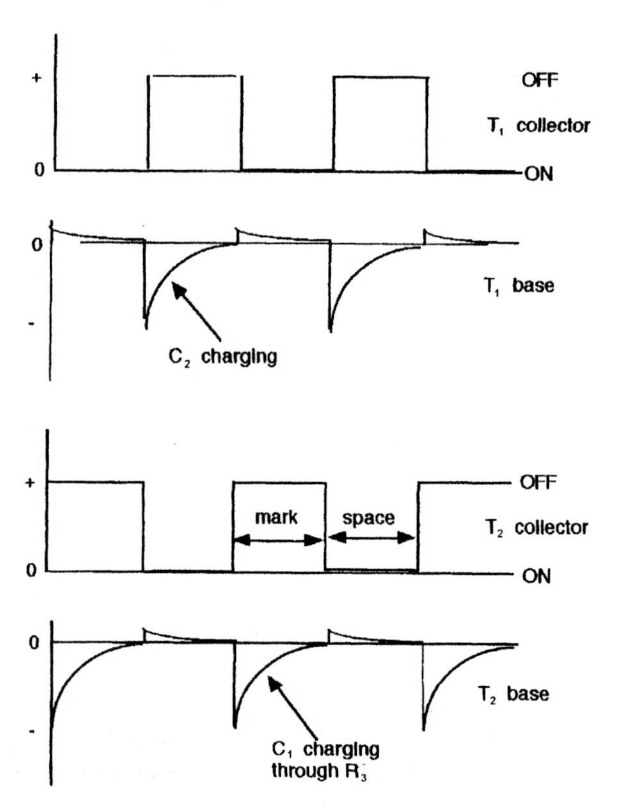

Figure 12.8 Voltage waveforms for an astable multivibrator

The bistable multivibrator

This circuit is very similar to the astable but it is not a free running oscillator. Each transistor conducts alternately as in the astable but to change which transistor is conducting a pulse must be provided from an external source. Figure 12.9 shows an example of a bistable oscillator.

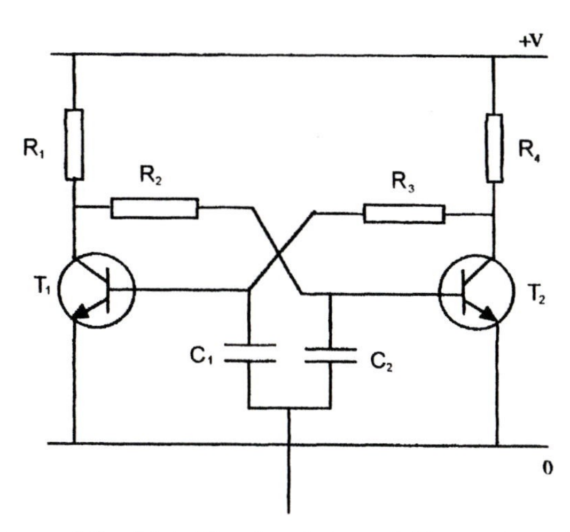

Figure 12.9 Bistable multivibrator

The operation of the bistable circuit is as follows. At switch on one of the transistors will conduct before the other just as in the astable. The base emitter bias for each of the transistors is controlled by the other so that if T_1 conducts and its collector voltage falls the base bias potential is removed from T_2 so its collector voltage is high which in turn ensures there is base bias for T_1. The bistable can stay in this condition as long as the supply is not interrupted or some other action is taken to change this state. If a negative pulse is applied to the base of T_1 it will turn off causing its collector voltage to rise. The result of a positive collector voltage on the collector of T_1 is to provide base bias potential through R_2 for T_2 which now turns on. The collector voltage of T_2 therefore falls and removes the base bias potential from T_1. The bistable has now taken up its other stable condition, hence its name *bistable,* a device with two stable conditions.

To improve the switching of the bistable it is desirable to ensure that the trigger pulse used to switch the state of the transistors is steered to the transistor that is conducting to switch it off. To do this the circuit is modified by the addition of a pair of steering diodes and biasing resistors and is shown in Figure 12.10.

To understand the operation of the steering diodes assume again that T_1 is ON and T_2 is OFF. The effect of this condition is that T_1 collector voltage is LOW and T_2 collector voltage is HIGH. The series circuit

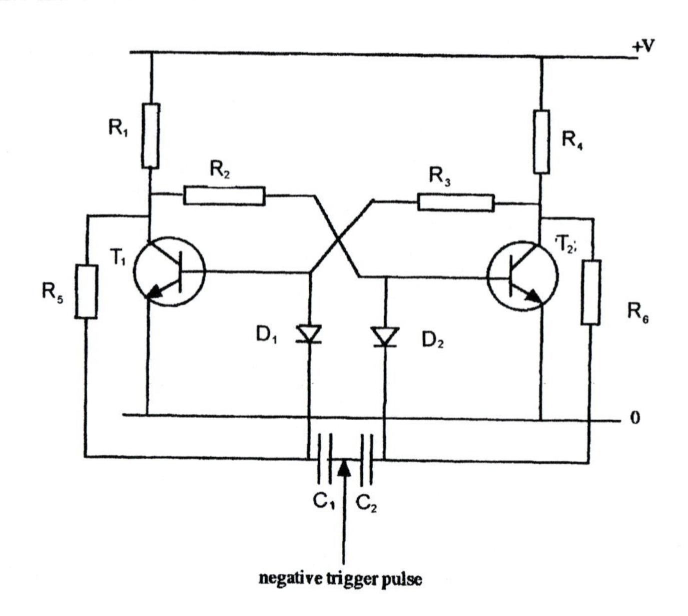

Figure 12.10 Bistable multivibrator with steering diodes

formed by R_6, D_2 and R_2 has D_2 reverse biased.The collector voltage of T_2 is high and is coupled via R_6 to the cathode of D_2 whose anode is connected via R_2 to the collector of T_1 which is low, hence D_2 is reverse biased. Examining the series circuit formed by R_5, D_1 and R_3 reveals that D_1 anode is positive since it is connected to the base of T_1 and has 0.7 V present while D_1 cathode is connected via R_5 to T_1 collector which is low because T_1 is conducting. D_1 although not forward biased (approximately 0.1 V on its cathode and 0.7 V on its anode) it is not held in reverse bias as is D_2. When a negative pulse is applied at the junction of C_1 and C_2 it easily forward biases D_1 and the pulse is steered to the base of T_1 to turn it off. On turning off T_1 the circuit automatically turns on T_2. The steering diodes are now biased in the opposite sense. D_1 is reverse biased while D_2 is almost forward biased so that when the next trigger pulse is applied it is steered to the base of T_2. The waveform produced by this circuit is shown in Figure 12.11.

The monostable multivibrator

This circuit is sometimes referred to as a *one shot* circuit, something that will become obvious as its operation is examined. The circuit diagram is shown in Figure 12.12.

On examining the circuit it is possible to see that T_1 base bias is directly coupled from the supply voltage through R_2 but T_2 gets its base bias through R_3 from the collector of T_1 and via R_1. At switch-on T_1 turns hard on causing its collector voltage to fall. This voltage

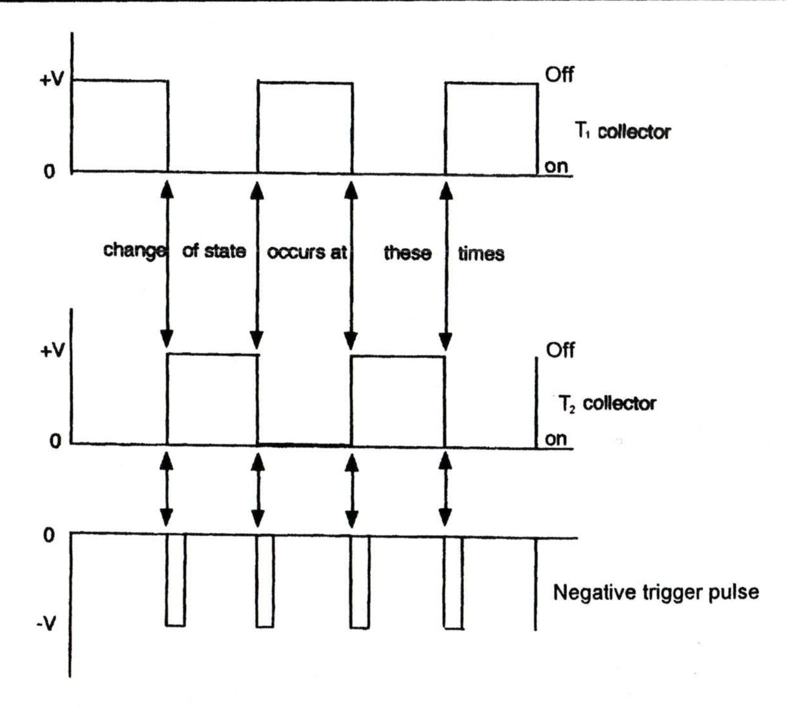

Figure 12.11
Bistable output and trigger waveforms

is not positive enough to provide any base bias for T_2 which therefore does not conduct. This is the stable condition for the monostable, a device with one stable condition. In order to cause a change in this condition a negative pulse is applied to the base of T_1 to turn it off. When T_1 is turned off its collector voltage rises and enables R_3 to provide the necessary base bias to turn T_2 fully on. The collector voltage of T_2 falls rapidly and in doing so charges the left-hand plate of C_1 negatively. This negative potential prevents T_1 conducting again even if the negative trigger pulse is no longer present. The monostable will remain in this condition therefore until the negative charge on C_1 can leak away through R_2 when the circuit reverts

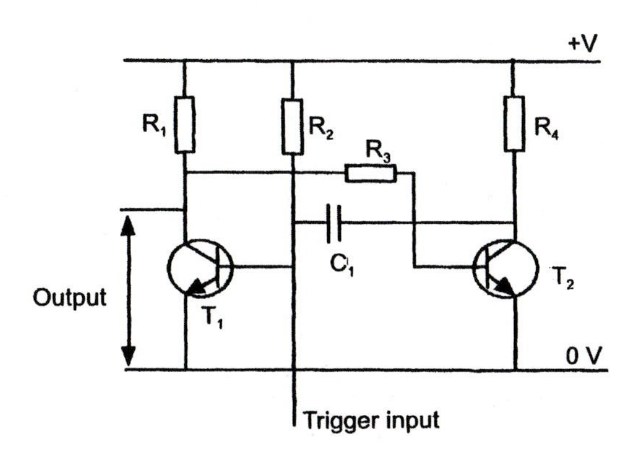

Figure 12.12
Monostable multivibrator

back to its normal condition with T_1 ON and T_2 OFF. The length of time the output pulse is present depends upon the value of the time constant formed by R_2 and C_1. The output waveform is shown in Figure 12.13.

Figure 12.13 Output and trigger pulse waveforms for a monostable

+V

0 V

C_1 charging through R_2

T_1 collector/output

0 V

Trigger pulses

–V

The Schmitt trigger

This is a form of bistable circuit where the output voltage condition is determined by the input voltage signal. Very often a squarewave becomes distorted during transmission and loses its sharply rising or falling edges. These edges are important since they define a specific point in time, so it is important to reshape the signal to redefine a precise switching point. This is illustrated in Figure 12.14.

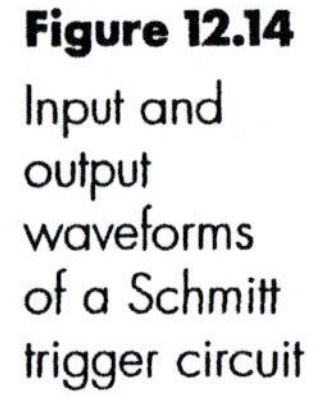

Figure 12.14 Input and output waveforms of a Schmitt trigger circuit

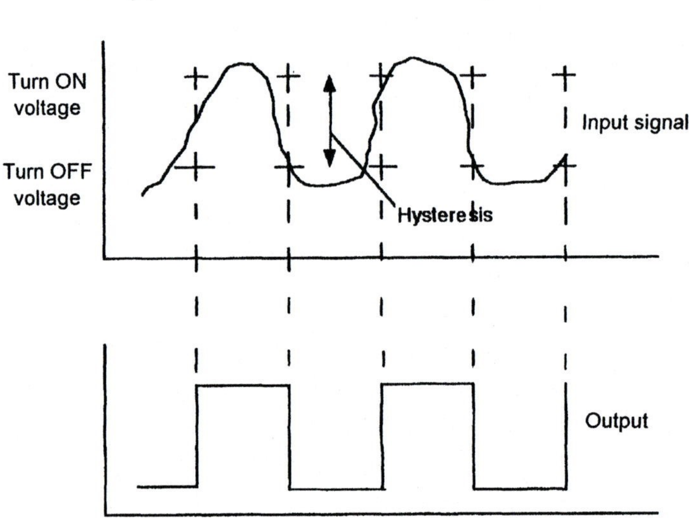

The circuit required to do this could be inside a logic gate or provided by a pair of transistors connected in the manner shown in

Figure 12.15. The circuit operates as follows. T_2 is forward biased by the resistor network R_1, R_2 and R_3 so it is conducting. The current flowing in T_2 produces an emitter voltage across R_4 which is a common emitter resistor for T_1 and since the voltage across R_4 is positive, T_1 emitter is more positive than its base, therefore T_1 is non-conducting. As the input voltage rises T_1 base becomes more positive than its emitter causing T_1 to conduct. The value of the collector load R_1 is relatively high so a small amount of collector current results in a substantial voltage drop in T_1 collector voltage which now reduces the base bias voltage for T_2. Since T_2 has less base bias the current flowing through it reduces so its emitter voltage developed across R_4 falls. T_1 emitter voltage must now reduce because both transistors share the same emitter resistor causing T_1 to conduct even more. Its collector voltage falls further and the cycle described continues very rapidly until T_2 is off. T_2 will remain off until the input signal at the base of T_1 falls to a level where the current flowing in T_1 reduces to a point where its collector voltage rises to a level sufficiently positive to allow T_2 to conduct again. When T_2 starts to conduct it rapidly turns off T_1 because of the shared emitter resistor. The difference between the turn-on and turn-off voltages is known as the *hysteresis* level.

Figure 12.15
Schmitt trigger

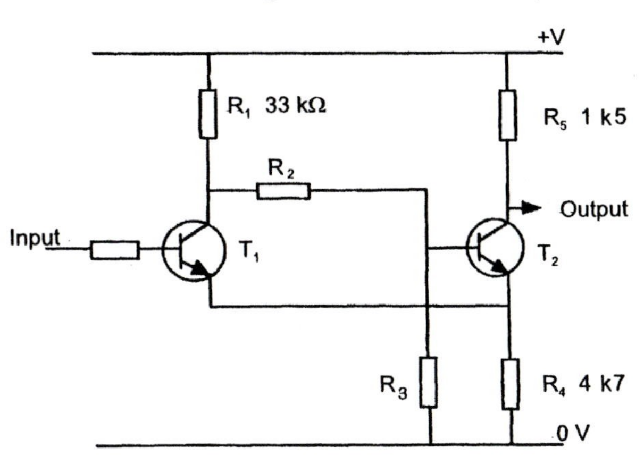

Operational amplifiers

This is not a special amplifier but describes a way in which a complex circuit conveniently packaged in an integrated circuit is connected and used. It was originally developed for use in analogue computers to perform mathematical functions. A schematic diagram is shown in Figure 12.16.

The amplifier shown in Figure 12.16 has a gain of *X* times. In other words the output is *X* times larger than the input. It also has an extremely high input impedance so it can be conveniently assumed that there is no current flowing into the amplifier. The only path that current can flow must therefore be through the input resistor R_1 and to the output via R_2 since they are both in series. If they are in series then clearly the current through both of the resistors must be the same so that $I_1 = I_2$. It has also been assumed that the amplifier introduces a phase shift or inversion of 180° exactly the same as a common emitter amplifier. Using simple Ohms law equations the following calculations reveal how it is possible to determine the actual gain of the amplifier.

Figure 12.16
Schematic circuit of an op amp

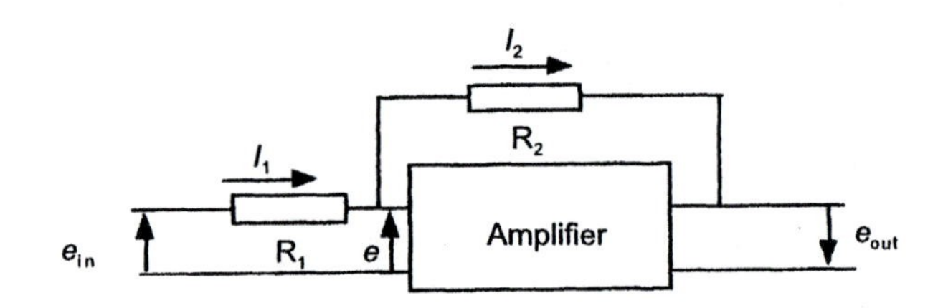

$$\text{The voltage across } R_1 = I_1 = \frac{e_{in}}{R_1}$$

$$\text{The voltage across } R_2 = I_2 = \frac{e_{out}}{R_2}$$

It has already been stated that $I_1 = I_2$ so the two equations must produce the same resultant.

$$I_1 = I_2 \quad \text{so} \quad \frac{e_{in}}{R_1} = \frac{e_{out}}{R_2} \quad \text{or} \quad \frac{R_2}{R_1} = \frac{e_{out}}{e_{in}}$$

The gain of an amplifier is determined by the signal voltage output divided by the signal voltage input so in the example used it is possible to determine the gain of the amplifier by examining the value of the resistors R_1 and R_2. If R_1 has the same value as R_2 the amplifier will have a gain of one or *unity,* but if R_1 is 10 kΩ and the value of R_2 is 100 kΩ the gain must be 10. By choosing the values of R_1 and R_2 it is possible to produce an amplifier with exactly the amount of gain required.

This is a very simplified description of an op amp just to identify its basic properties, but in reality it is a complex circuit composed of a large number of high gain amplifiers directly coupled together using feedback. The op amp is normally used with negative feedback because without it the *open loop gain* is far too large to be useful, typically 100 000 or more so a feedback resistor R_2 is used to establish the desired gain.

Properties of an op amp

An ideal op amp will have the following properties:

(i) very high input impedance
(ii) very low output impedance
(iii) very high voltage gain in open loop mode
(iv) very wide bandwidth

A commonly available and useful op amp is a device known as a 741. Using this device as an example, its characteristics are:

Input impedance 1 MΩ
Output impedance 150 Ω
Open loop gain 100 dBs
Bandwidth 1 kHz
Maximum supply voltage +/−18 V, it needs a dual supply
Maximum load current 10 mA

Although the bandwidth quoted for this device appears to be low it has been in use for many years and was actually designed for low frequency operation. The gain/frequency characteristic of an op amp is shown in Figure 12.17.

Figure 12.17 Frequency response of an op amp

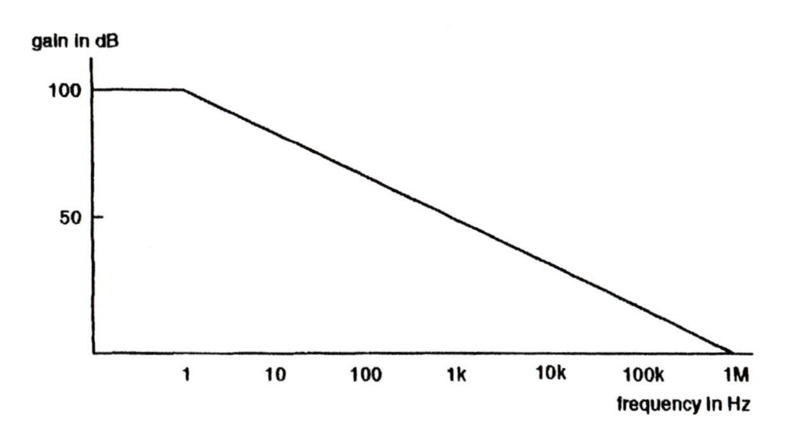

The gain of an amplifier and its bandwidth are related and there is always a gain/bandwidth product so that increasing the amount of negative feedback *reduces* the *gain* but *increases* the *bandwidth.* The circuit designer therefore has the opportunity to produce with accuracy the amplifier desired. Figure 12.18 shows the circuit symbol and pin lead out connections for a 741 IC, along with a simple circuit arrangement.

The op amp is connected as shown in Figure 12.18 with its non-inverting input connected to ground through R_2 and its inverting input to the incoming signal through R_{in}. R_f is the feedback resistor.

The output will be the result of amplifying the signals at both the inputs but as the non-inverting input has no signal on it the output produced must be an inverted version of the signal at the inverting input. The amplitude of the signal will be determined by the ratio of the value of the resistors R_f and R_{in} as stated previously. This makes an assumption that if there is no signal present at both inputs or an equal signal at both inputs the amplifier will produce no output. Unfortunately this does not always occur because of temperature changes and it is necessary to incorporate some compensation. This is the purpose of the *offset null* connections which balance the circuit. R_2 helps minimise the effects of *drift* caused by temperature changes and its value is made equal to the parallel combination of R_f and R_{in}.

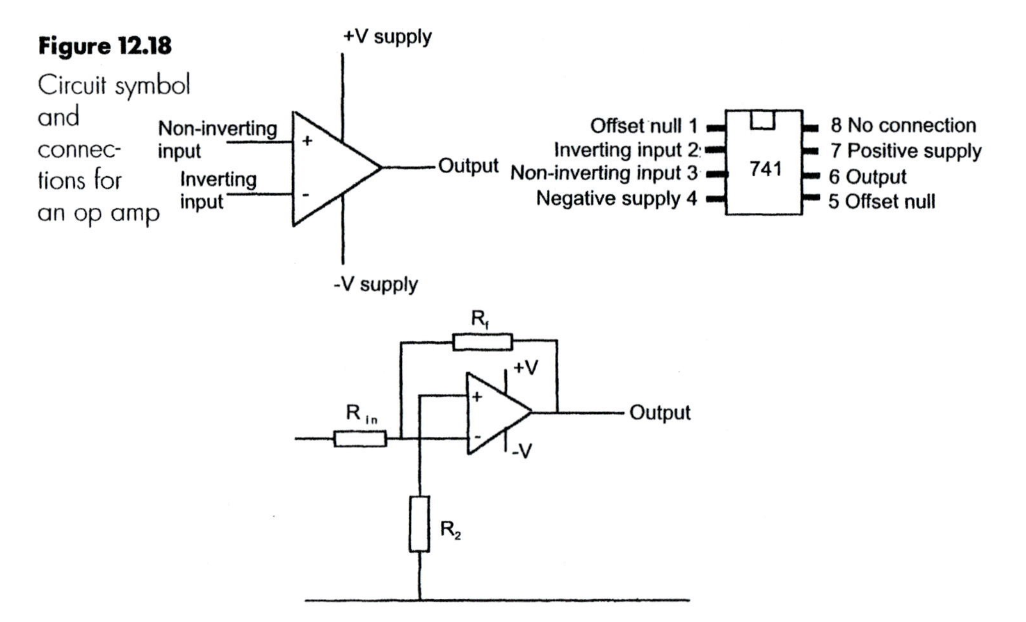

Figure 12.18 Circuit symbol and connections for an op amp

Summing amplifier

One of the most useful purposes of the op amp is as a summing amplifier because it can have a number of signal sources connected to one of its inputs. The circuit shown in Figure 12.19 shows such a circuit with four inputs A, B, C and D. Although only one amplifier is used it is possible to deal with each input as if it was the only input connected to the amplifier when determining the amount of amplification given to its signal.

The gain from input A is determined by the ratio of the resistors R_1 and R_f while the gain from input B is determined by the ratio of the resistors R_2 and R_f and similarly for the remaining inputs. Simple

Figure 12.19
Simplified circuit of an op amp summing amplifier

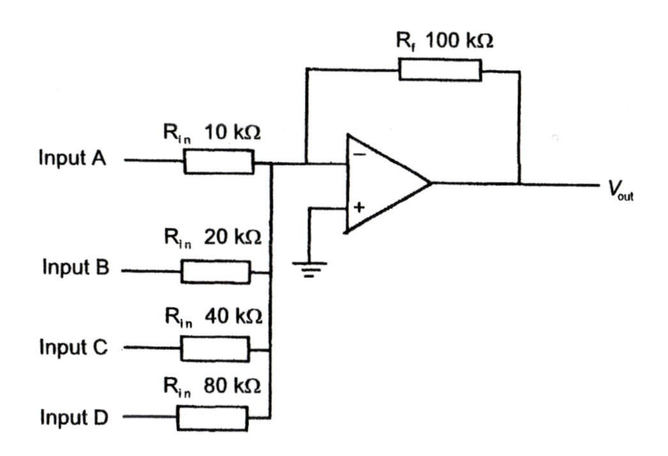

values have been chosen to illustrate the point but in practice the values will be much different. Suppose the input signal to each input is 1 volt, again to keep the calculation simple. The output produced by each of the inputs will be as follows:

$$\text{Gain} = \frac{R_f}{R_{in}} \quad \text{or} \quad V_{out} = V_{in} \times \text{gain}$$

alternatively this simplifies to $V_{out} = \dfrac{V_{in} \times R_f}{R_{in}} = \text{volts}$

For input A the output would be equal to 10 volts since the ratio of the resistors R_{in1} and R_f is 10 kΩ to 100 kΩ or 1:10 so the amplifier has a gain of 10. This can of course be proved using the given formula. The output produced by the signal applied at input B would not be the same amount as at input A because the ratio of R_{in} to R_f is now 20 kΩ to 100 kΩ or 1:5; the gain for this input is only 5 hence the output produced by this input would be 5 volts or half the amount produced by input A when the same amplitude signal is applied. This same argument can be applied to inputs C and D using the ratio of the values of R_{in} to R_f to determine the gain for each. The total output available from the amplifier will be the sum of all of the voltages produced by the input signals and the amount of gain the amplifier gives to each input. For the example used where each input has a 1 volt signal applied and the resistor values shown in Figure 12.19 the output can be calculated as follows:

$$V_{out} = \left(\frac{V_{in} \times R_f}{R_{in1}}\right) + \left(\frac{V_{in} \times R_f}{R_{in2}}\right) + \left(\frac{V_{in} \times R_f}{R_{in3}}\right) + \left(\frac{V_{in} \times R_f}{R_{in4}}\right)$$

$$= \left(\frac{1 \times 100\ \text{k}\Omega}{10\ \text{k}\Omega}\right) + \left(\frac{1 \times 100\ \text{k}\Omega}{20\ \text{k}\Omega}\right) + \left(\frac{1 \times 100\ \text{k}\Omega}{40\ \text{k}\Omega}\right)$$

$$+\left(\frac{1 \times 100\,\text{k}\Omega}{80\,\text{k}\Omega}\right)$$

$$= 10 + 5 + 2.5 + 1.25$$

$$= 18.75 \text{ volts}$$

This would of course be a negative voltage since a positive voltage has been applied to the inverting input. It must be stressed again that these figures have only been used for convenience to illustrate how the circuit functions. The importance of this particular summing amplifier is that it makes an ideal circuit to use for converting a 4 bit digital signal into an analogue value since the value of output produced by input A is twice that of input B while the output produced by input B is twice that of input C and so on, just as a binary value. The op amp has many other uses, audio mixers, instrumentation and measuring equipment, etc. It is used because it is possible to build a device easily that behaves in the desired manner.

The transistor as an oscillator

An electronic oscillator is a device which can convert the dc supply voltage into an alternating output. This could be either as a squarewave (described earlier in this chapter) or as a sine wave. Having already dealt with a few types of squarewave oscillator it is now important to understand the principles of a sine wave oscillator.

Sinusoidal oscillators fall into two groups:

(i) L–C oscillators – These use tuned circuits to determine their operating frequency.
(ii) R–C oscillators – These use a combination of resistors and capacitors to determine their operating frequency.

However, both of these types of oscillator rely on *positive feedback* to enable them to function.

Positive feedback

This effect occurs when part of the output signal of an amplifier is coupled back to the input. A typical example being when a microphone is placed next to the loudspeaker in an public address system to produce a howling noise. This is positive feedback. Figure 12.20 shows the block diagram of a sinusoidal oscillator.

L–C oscillators

Figure 12.21 shows a parallel tuned circuit. Capacitor C_1 has been charged from a dc supply with the polarity shown. If S_1 is closed electrons from plate B pass through L_1 to plate A attracted by its positive charge. This of course creates a magnetic field around the inductance L_1. When the excess of electrons on plate B have passed to plate A the magnetic field around this inductance collapses and in doing so charges plate A of the capacitor negatively. The excess of electrons on plate A are now attracted by the positive charge on plate B and pass through the inductance L_1 again but this time in the opposite direction. A magnetic field is established around the coil again but it has the opposite polarity because the current is flowing in the opposite direction. Plate A loses its excess electrons, the magnetic field around L_1 collapses and plate B is recharged negatively. The circuit is back to the starting condition, well almost. The circuit has resistance so energy is lost each time the cycle repeats and eventually it will all be dissipated so the oscillation ceases, rather like a weight on a piece of string that is allowed to swing. This is called a *damped oscillation*. To maintain the oscillation shown in Figure 12.22 a method has to be found of replacing the lost energy, recharging the capacitor to maximum at point A. The amplifier serves this function. The frequency of oscillation is determined by the value of capacitor and inductor used in the tuned circuit and can be calculated by using the formula:

$$f = \frac{1}{2\pi\sqrt{LC}}$$

where L = inductance in henries, C = capacitance in farads and $\pi = 3.14$.

Figure 12.20
Block diagram of a sinusoidal oscillator

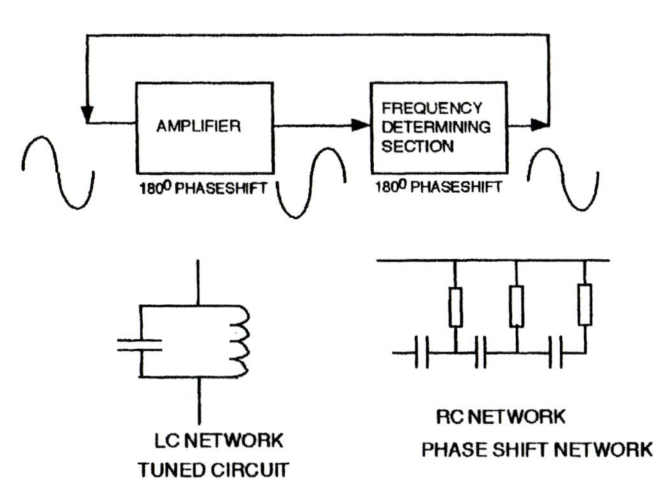

This circuit can be thought of as an energy shuttle. The energy starts off in the electrostatic field of the capacitor and as the capacitor discharges it is transferred to the electromagnetic field around the inductor and back again.

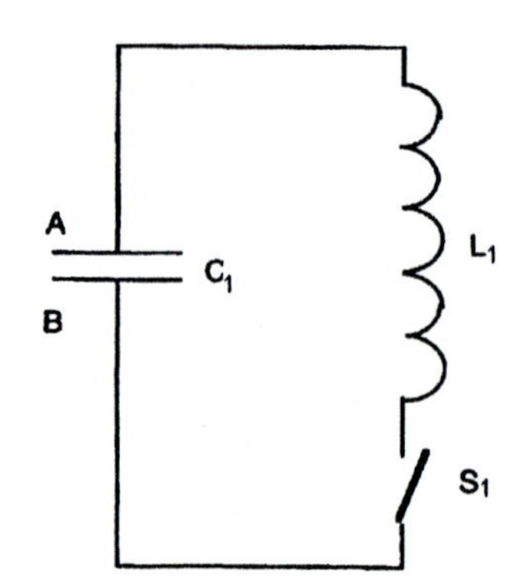

Figure 12.21
Tuned circuit

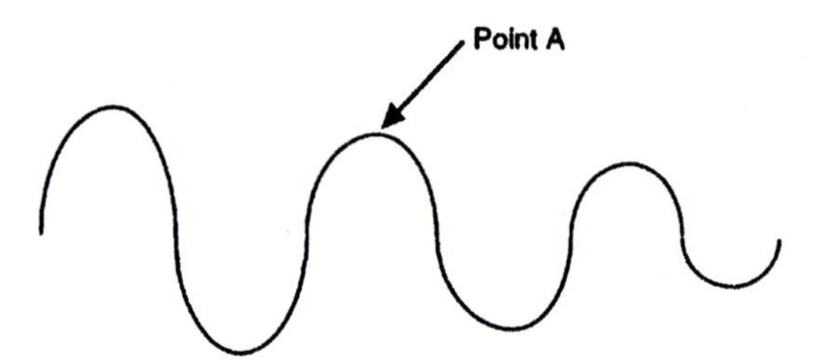

Figure 12.22
Damped oscillation

The Hartley oscillator

The circuit diagram of a Hartley oscillator is shown in Figure 12.23. This is a well-known and much-used circuit which can be identified by the use of an inductance with a centre tap.

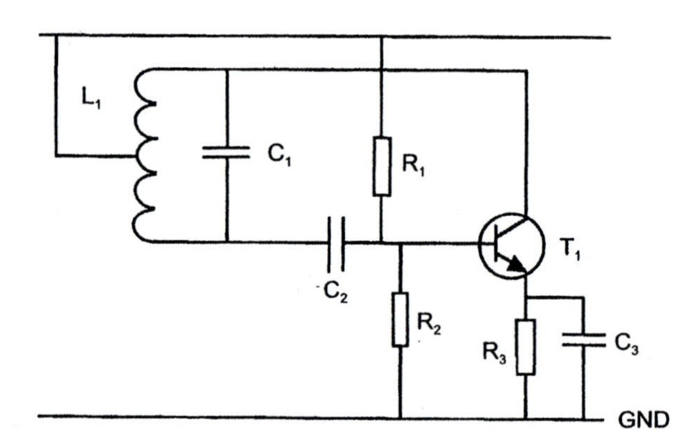

Figure 12.23
The Hartley oscillator

The operation of the circuit is relatively simple to follow. To start oscillating T_1 needs its base/emitter junction to be forward biased as it did in the common emitter amplifier. This is logical since the transistor acts as an amplifier in this circuit. To obtain positive feedback the signal at the base of T_1 is inverted by normal transistor action achieving 180° of phase difference and then by the action of the inductance L_1 a further phase difference of 180° making a total phase difference of 360° so that the feedback is in phase with the original input. R_3 and C_3 provide a *self-biasing network* although they appear to be an emitter stabilising resistor and bypass capacitor found in the common emitter amplifier.

The Hartley oscillator relies on the use of a centre tapped coil as shown in Figure 12.24. The action of the coil is to produce a further 180° of phase shift so that positive feedback occurs. When the signal at the top of the coil X is positive the signal at the bottom of the coil Y must be negative because of self-inductance. C_2 serves simply as a dc blocking capacitor and prevents the higher dc collector voltage arriving at the base of T_1 whilst allowing the ac signal to pass. The frequency of oscillation is determined by C_1/L_1.

Figure 12.24
Phase relationship across centre tapped coil

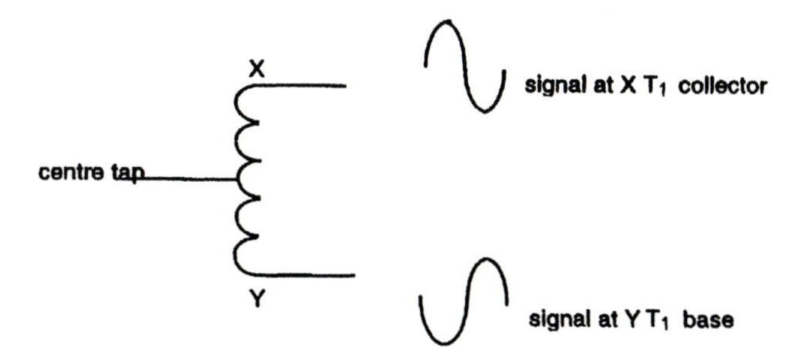

At switch-on T_1 conducts. The base must be positive (NPN transistor) and a rise in base voltage causes a fall in collector voltage. By the action of the inductor producing positive feedback the base voltage rises quite quickly since the signal fed back adds to the dc starting bias. Eventually the voltage on the collector cannot fall any further so the positive feedback falls. The collector voltage rises until it reaches maximum and when the current ceases to change the magnetic field in the coil ceases and the device is back to its staring point and the cycle repeats. The rate of charge and discharge of C_1 and L_1 determine the frequency of operation.

When the oscillator is operational the dc voltage applied to the base of T_1 to start it conducting produces an emitter current through R_3. The voltage across R_3 as a result of this current charges capacitor C_3 and when the feedback signal is added to the base bias voltage the charge on C_3 becomes only slightly less than the maximum positive value on the base. R_3 and C_3 also have a time constant. When the transistor turns off the only path for the capacitor to discharge is through R_3. If the time constant of R_3/C_3 is chosen so that it is long when compared to the periodic time of the oscillation the emitter voltage will still be positive when the signal at the base is negative, which of course turns off the transistor. Once the circuit commences its operation the biasing moves from what is described as *class A* into *class C*. The transistor only turns on when a positive peak of signal is present at the base to restore the lost energy and maintain the oscillation. The change from class A to class C bias can be very useful from a service engineer's point of view since it is a simple method of determining if an oscillator is working without the use of an oscilloscope. Under normal conditions the voltage on the base of an NPN transistor will be 0.6V higher than the emitter (when in class A). However, if the device is oscillating it moves to a position where the voltage on the emitter may be slightly more positive than that of the base. This can only be true if the circuit is oscillating, so a simple voltmeter check can be used. The use of R_3/C_3 is also advantageous because it gives the oscillator amplitude stability since the voltage on the emitter can only change at a rate determined by the time constant of R_3/C_3. If the transistor becomes warm and produces more output the voltage on the emitter rises and decreases the amount of forward bias which reduces the output amplitude. Similarly if the supply voltage rises or falls the charge on C_3 changes accordingly to maintain the output amplitude.

Phase shift oscillators

Just as the *LC* oscillator had two sections, an amplifier or maintaining section and a frequency determined section also producing phase shift, so too does the *RC* oscillator. Its circuit diagram is shown in Figure 12.25.

Figure 12.25
RC phase shift oscillator

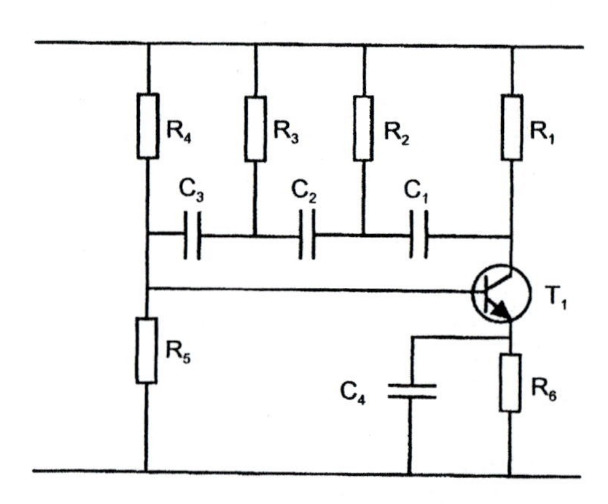

In Figure 12.25 T_1 acts as an amplifier and therefore produces a phase difference of 180° between base (input) and collector (output). The phase shift and frequency determining section is made up of the 'ladder network' R_1, C_1, R_2, C_2, R_3 and C_3. Each combination of *R* and *C* provides 60° of phase shift which can be explained by the use of a phasor or vector diagram.

This is shown in Figure 12.26(a) to (d). The supply current I_s is common to both R_1 and C_1 and can be considered as the reference (Figure 12.26(b)). The voltage V_r across R_1 must be in phase with I_s since the voltage and current are in phase in a resistor (Figure 12.26(c)). The voltage and current in a capacitor are not in phase with the voltage V_c lagging the current Is by 90°. The relationship of these voltages and the current is shown in Figure 12.26(d).

Figure 12.26
Phasor diagrams for an *RC* network

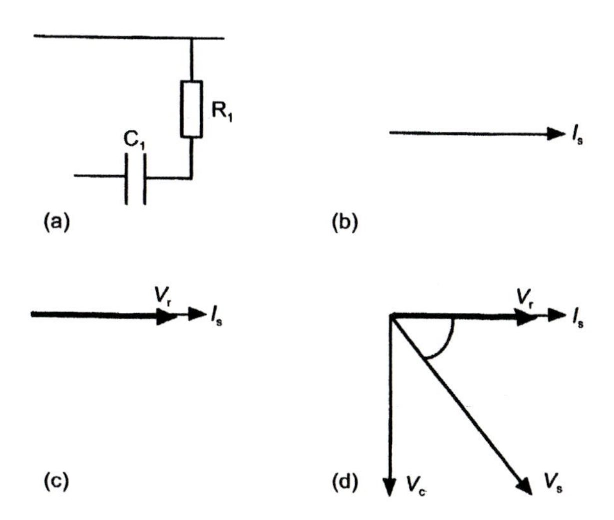

The phasor diagram enables the phase of the resultant voltage V_s to be shown and reveals that a phase shift of 60° has occurred. Since the phase shift oscillator has three *RC* networks the result is a phase shift total of 180°. The signal fed back is now in the same phase as the input so positive feedback has been achieved and oscillation will occur. The frequency of operation is determined by the time constant produced by the values of *R* and *C*. Different values of *R* and *C* give different time constants and therefore different frequencies of oscillation. The role of the transistor is important since it must provide enough gain to maintain oscillation. Experimentation shows that a minimum gain of 29 is needed. The frequency of operation can be determined using the following formula:

$$f = \frac{1}{2\pi\sqrt{6}RC}$$

Other types of phase shift oscillator are possible notably the Wein bridge oscillator. Its block diagram is shown in Figure 12.27.

Phase shift oscillators are normally used at low frequencies usually as audio frequency oscillators. In the case of the phase shift oscillator using a ladder network of *R* and *C* this would be a fixed frequency oscillator since it would be difficult to arrange a mechanism to adjust the components necessary in all three sections of the *RC* ladder at the same time. However, in the Wein bridge oscillator varying the frequency is simpler since only two resistors and two capacitors are used in the phase shift network. To change the operating frequency a dual variable resistor or a ganged twin tuning capacitor can be used.

Figure 12.27
Wein bridge phase shift oscillator

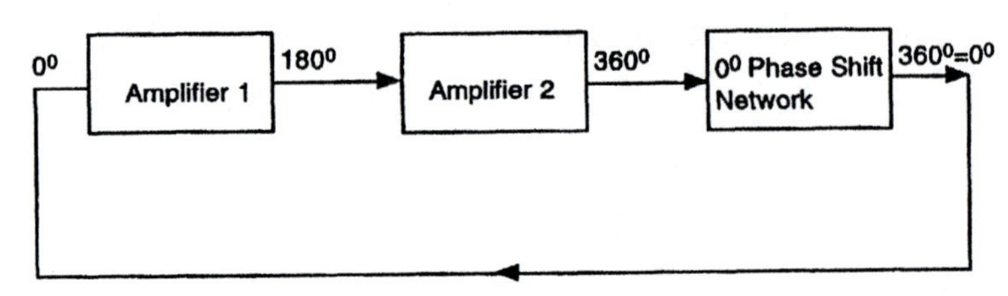

Block diagram of a Wein Bridge Phase Shift oscillator

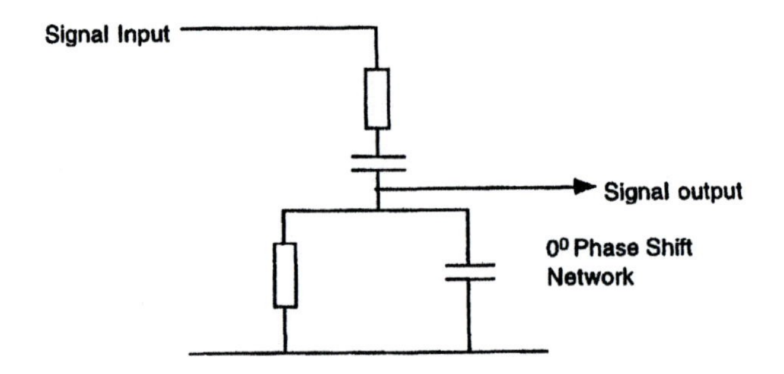

13

Electronic systems

Electronic devices are constructed by connecting together various circuits which perform their own special function. For example, a radio receiver contains an oscillator, a mixer, a set of tuned IF amplifiers, a demodulator and an audio amplifier. Quite a complex arrangement. Each of these devices requires different circuits, devices and signals.

DC signals

A direct current, i.e. the type supplied by a battery, has only limited use in conveying information. A simple circuit using dc signalling is shown in Figure 13.1.

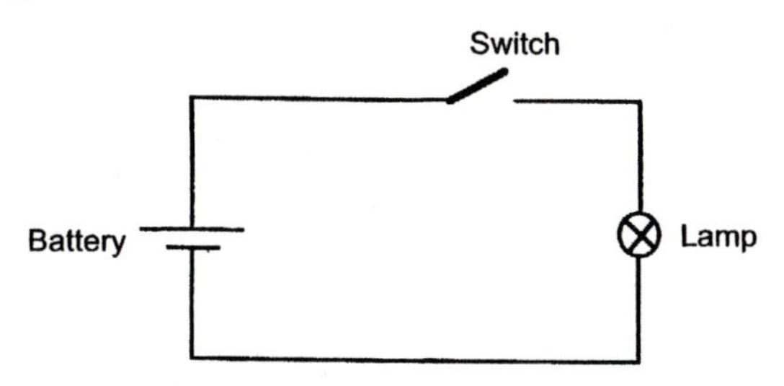

Figure 13.1
Simple dc signalling system

When the switch is closed current flows in the circuit which causes the bulb to light. This indicates a condition has arisen, but the information is limited. The amount of information conveyed could be increased by using a code such as Morse Code. In the circuit of Figure 13.1 current flows only one direction when the switch is closed. It is said to be a *unidirectional* current. If the battery connections are reversed current flows in the opposite direction. If some method of switching can be arranged it is possible to create a current flow which is *bidirectional.* In the example in Figure 13.1, the lamp circuit, changing the polarity of the battery will make no real difference since the bulb will light whenever current flows through it, but if a small dc motor is used in place of the bulb its direction of rotation will change. This system is used in many model train sets, but again only a simple command system exists to convey a small amount of information, forward, reverse, fast and slow.

Speech is obviously the best way of communicating person to person, but unaided the distance over which people can communicate is very limited. Many different methods of communication have been tried and used over the years.

In Figure 13.2 a simple telegraphy system is shown. This uses a bidirectional system since current must flow in both directions. It also uses a coding system but this makes the system limited by the connecting wires. The greater the length of the connecting wires the greater the voltage dropped. To enable communications over long distances using telegraphy it is necessary to have intermediate stages which of course makes the system more complex. Telegraphy in its simple form as shown in Figure 13.2 cannot be used to convey speech or music but it led to the development of a telephone system which uses ac signals.

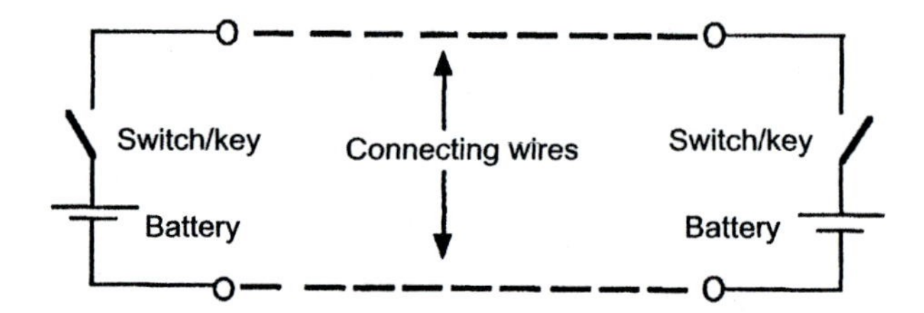

Figure 13.2
Simple telegraphy system

Alternating current signalling systems

To develop an ac signalling system it is important to understand the nature of an ac signal. Sound produced by a human voice or a musical instrument is the result of variations of air pressure which travel through the atmosphere and are picked up by the ears. Taking the example of a tuning fork, when it is struck its prongs vibrate and cause the air around it to be alternately compressed and rarefied. It is these variations in air pressure which are sensed by the ears, and the sound waves cause the ear drums to move in a similar manner to the movement of the tuning fork prongs. So sound is heard. It has to be emphasised that the air does not move but the pressure *alternates* in the manner shown in Figure 13.3.

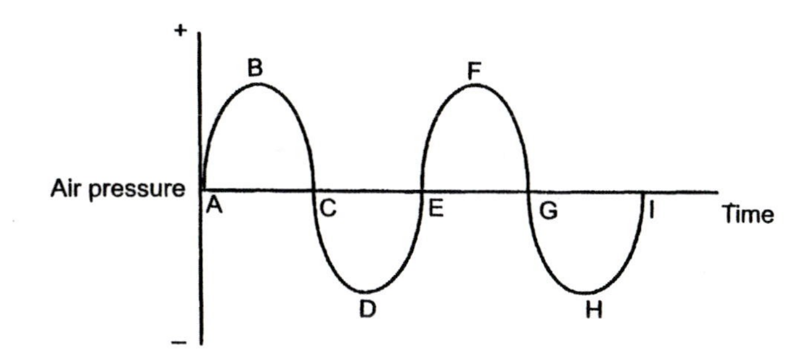

Figure 13.3 Sinusoidal waveform or sine wave

From one point on this waveform to the next corresponding point is identified as a complete cycle. Hence from point A to point E, or point B to point F, or point D to point H, all represent one complete cycle. The number of cycles completed in one second is called the *frequency* and is measured in *hertz.*

One cycle in one second = 1 hertz or 1 Hz

The frequency of a sound wave (the number of complete cycles per second) determines the pitch. So a low pitched note has a low frequency and a high pitched note a high frequency. The range of frequencies which the ear can respond to is limited and the extremes between 20 Hz and 20 kHz are called the *audio frequency range (AF).* The range of frequencies within this that each individual can hear varies. If the sound becomes louder the variations in air pressure become greater.

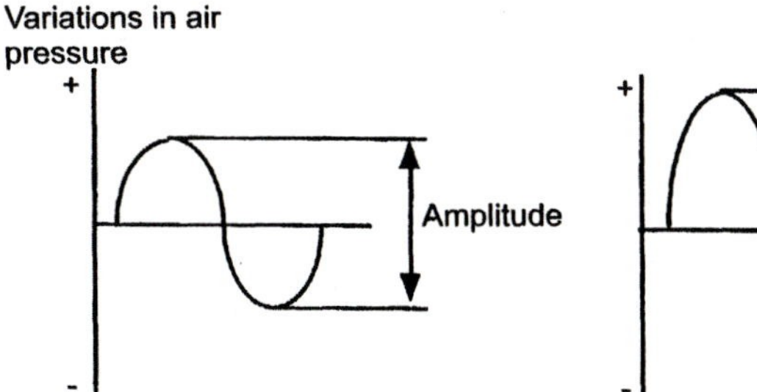

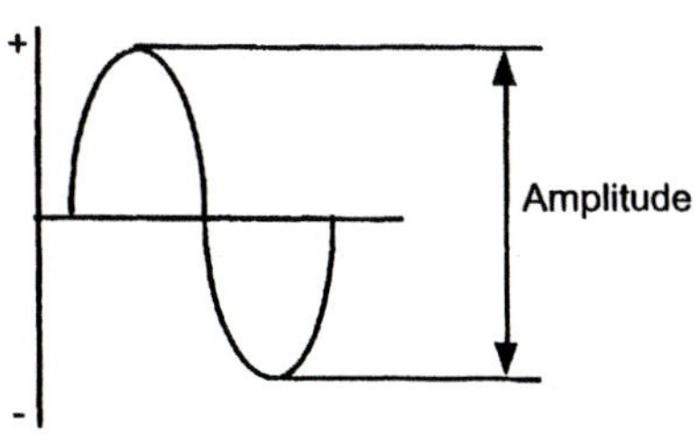

Figure 13.4 Effect of increasing the variations in air pressure

Speech and music produce waveforms which, when examined, are complex and do not appear to be sinusoidal. However, the shape of any waveform depends upon its source and it is this that enables us to distinguish one person's voice from another person's voice and one musical instrument from another. Although the waveforms produced do not appear to be sinusoidal careful examination has shown that they are the product of a number of sine waves added together. The waveform produced by a musical instrument therefore consists of a sine wave at the base frequency known as the *fundamental* frequency and others which are multiples of the fundamental frequency and are called *harmonics.* A harmonic at twice the frequency of the fundamental is called the second harmonic, while a harmonic at three times the frequency of the fundamental is called the third harmonic.

For example, a non-sinusoidal waveform has a frequency of 500 Hz. It contains frequencies which on examination are found to be the second harmonic and the third harmonic. Identifying the content of this waveform reveals the following frequencies to be present:

Fundamental	=	500 Hz		
Second harmonic	=	fundamental × 2	=	500 Hz × 2
	=	1 kHz		(1000 Hz)
Third harmonic	=	fundamental × 3	=	500 Hz × 3
	=	1.5 kHz		(1500 Hz)

Consider a non-sinusoidal wave of 6 kHz (6000 Hz). It contains a range of frequencies up to and including the fifth harmonic. Its frequency content is therefore sine waves at:

Fundamental frequency	6 kHz	
Second harmonic	12 kHz	fundamental × 2
Third harmonic	18 kHz	fundamental × 3
Fourth	24 kHz	fundamental × 4
Fifth	30 kHz	fundamental × 5

If this particular signal is to be transmitted or amplified it is important that all the harmonics are amplified equally to reproduce exactly the same shaped signal. If this is not done distortion will be introduced and although the frequencies of the fourth and fifth harmonic cannot be heard their presence is important. This is the reason why high quality hi-fi (high fidelity) amplifiers have *bandwidths* of between 5 Hz and 100 kHz. If harmonic frequencies are removed the reproduced sound will be different, a typical example being the telephone system. Individuals do not sound the same via the telephone network as they do face to face because some of the harmonics are removed.

As can be seen in Figures 13.5 and 13.6 the shape of the resultant waveforms is not sinusoidal. The effect of adding the fundamental to the harmonic produces a resultant that is less sinusoidal in appearance. Its shape is dependent on the relative *amplitudes* and *phases* of the signals. The phase of a signal is the point at which it commences its cycle. Figure 13.7 shows the effect of two sine waves at the same frequency but in different phases.

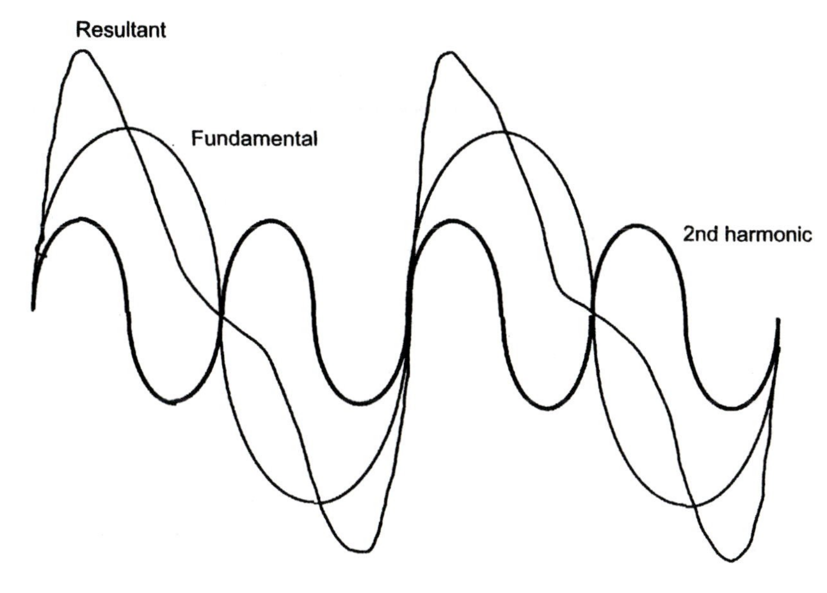

Figure 13.5
Effect of adding the second harmonic to the fundamental

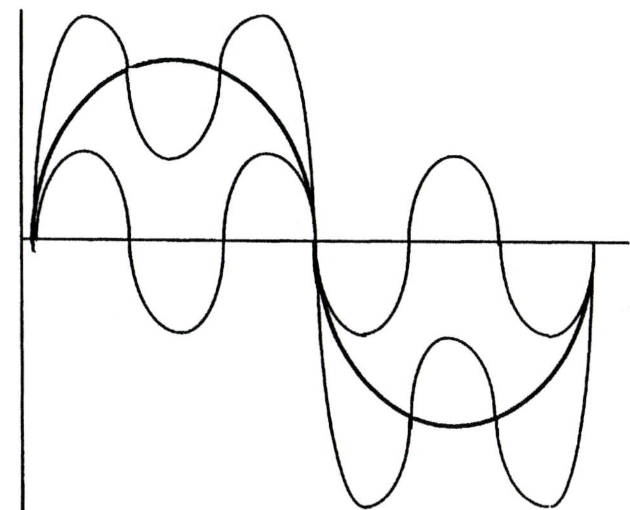

Figure 13.6
Effect of adding the third harmonic to the fundamental

Changes in phase might change the shape of the resultant waveform but should not affect the reproduced sound since the human ear is not able to detect changes in phase.

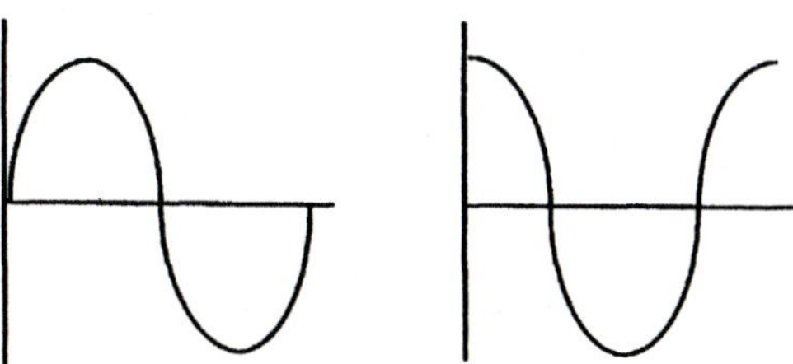

Figure 13.7
Out of phase sine waves

To transmit changes in air pressure or to amplify them it is first of all necessary to convert the variations of air pressure into electrical signals which are varying voltages and currents. Devices which convert one form of energy into another form of energy are called *transducers*. Examples of transducers used in an audio system are:

1. The microphone – Converts sound waves (varying air pressure) into voltage/current.
2. The loudspeaker – Converts voltage/current variations into sound waves.

To complete the audio system an amplifier has to be included to produce the power required, so a simple audio system has the devices shown in Figure 13.8 to carry out the functions identified.

Figure 13.8 Simple audio system

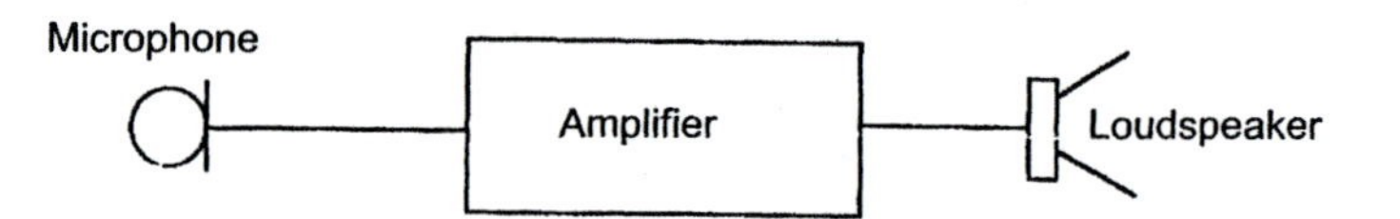

For the reproduced sound to be a larger but faithful copy of the original signal (a copy free of distortion), each part of the system must be able to pass not just the fundamental frequency but the harmonics as well. The simple sound system of Figure 13.8 has been chosen to illustrate this point. When constructing a system the design engineer must take into account that an ac signal at whatever frequency may well not be a single frequency but a fundamental frequency and its harmonics. Removing harmonics, changing their amplitude or phase can have a serious effect, hence it is important to ensure that each part of the system has the capability of handling the appropriate range of frequencies. This simple audio system is the basis of a telephone system. Sound waves are converted by the use of a suitable transducer into an electrical signal, amplified and then passed to another transducer to convert the electrical signals back into sound waves. In the case of a telephone system each telephone must have a microphone, amplifier and loudspeaker (or equivalent), whilst a complete telephone system has a complex network of switches and wires to interconnect each telephone to another telephone as required. It is also worth mentioning at this point that the range of frequencies a system can operate across is known as the *bandwidth*. There is a standard form of measurement to determine the bandwidth of a system or device. The maximum output is measured and the range of frequencies over which the system can produce an output within 3 dB of this maximum amplitude is termed the bandwidth. Remembering what has been said already, that removing harmonics or changing their amplitude or phase causes distortion, it is perhaps a little clearer why a person sounds different when they speak through the telephone network than when they speak directly face to face. The range of frequencies passed by the telephone network is restricted, for a number of reasons, to a small section of the AF range, 300 Hz–3.5 kHz, which removes many of the harmonics produced by a normal human voice, hence the effect is a voice, which sounds a little different.

Frequency/time

Frequency and time are related since frequency is said to be the number of complete cycles in one second. The time taken for one cycle must therefore be one second divided by the number of complete cycles per second. This time is called the *periodic time* and given the symbol t:

$$t = \frac{1}{f}$$

Where f = frequency.

The periodic time of a 10 kHz sine wave is:

$$t = \frac{1}{f} = \frac{1}{10\ \text{kHz}} = \frac{1}{10 \times 10^3} = 0.1 \times 10^{-3} = 0.1\ \text{ms} \quad \text{OR } 100\ \mu\text{s}$$

Squarewaves

A symmetrical squarewave as shown in Figure 13.9 can be proved to be the result of the addition of a sine wave at the fundamental frequency and its *odd* harmonics. The amplitude of the harmonic decreases inversely proportional to its harmonic number, for example the amplitude of the third harmonic will be $\frac{2}{3}\pi$ of the fundamental's amplitude and the amplitude of the fifth harmonic will be $\frac{2}{5}\pi$ of the fundamental's amplitude. Put in simple terms, this can be summarised by saying as the frequency of the harmonic increases its amplitude becomes smaller.

Figure 13.9
Symmetrical squarewave

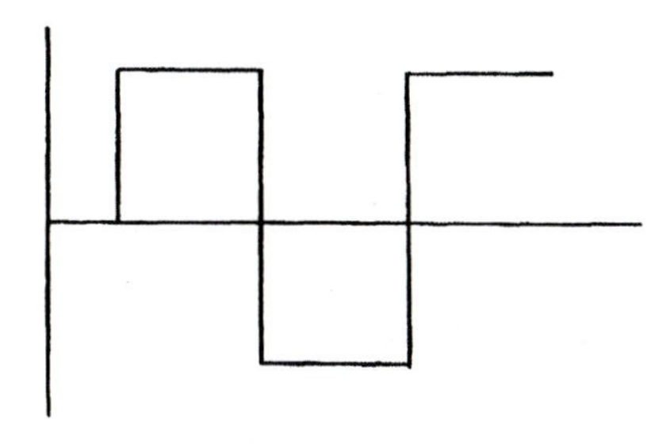

The asymmetrical squarewave as shown in Figure 13.10 is the resultant of adding a fundamental frequency to a wide range of its harmonics. It can best be considered as a series of positive pulses repeated at regular intervals. The number of pulses occurring per second is known as the *pulse repetition frequency* and the time occupied by one pulse is known as the *pulse duration*.

Figure 13.10
Asymmetrical squarewave

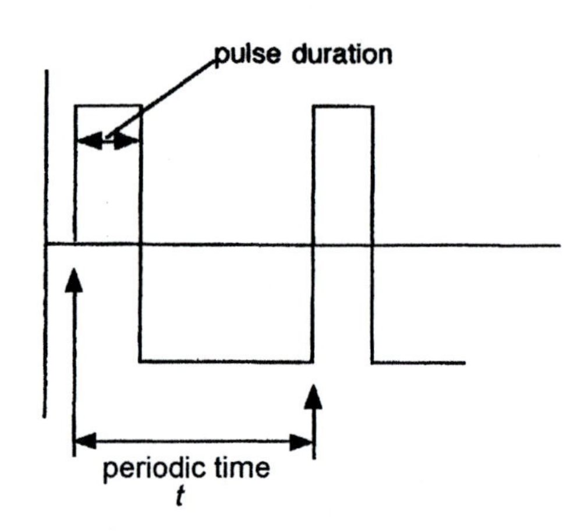

Rise and fall time

In practice the change from minimum to maximum (or the reverse) amplitude may not occur instantaneously, hence the leading and trailing edges of the pulse will not be vertical. The pulse may take a period of time to change from minimum to maximum (or the reverse). The time taken by the pulse to rise from 10% to 90% of its maximum value is called the *rise time*, while the time taken by the pulse to fall from 90% to 10% of its maximum value is called the *fall time*. This is shown in Figure 13.11.

Figure 13.11
Rise and fall time measurement

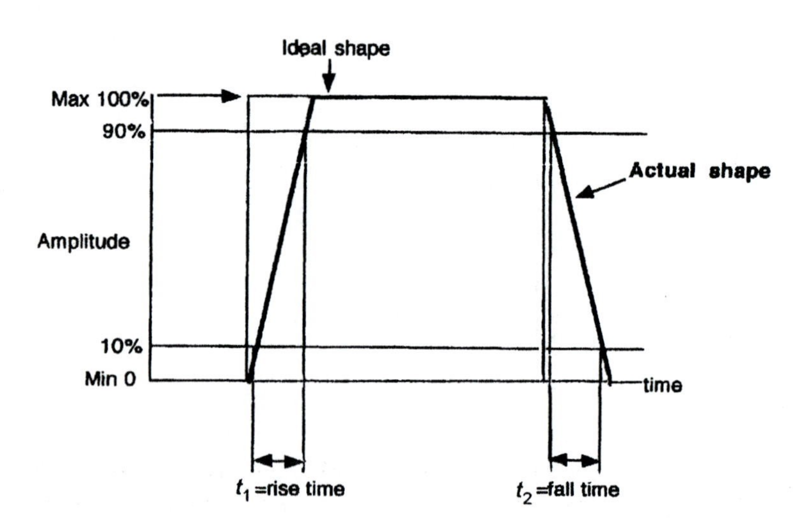

Mark to space ratio

The ratio of the pulse duration to the interval between pulses is known as the mark to space ratio. The example shown in Figure 13.12(a) has a ratio of 4 to 1 since the time the pulse is positive is four times longer than the time the pulse is negative. If this situation is reversed the mark to space ratio becomes 1 to 4 since the pulse is now only positive for a quarter of the time it is negative and is shown in Figure 13.12(b).

Figure 13.12

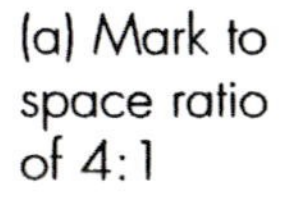
(a) Mark to space ratio of 4:1

(b) Mark to space ratio of 1:4

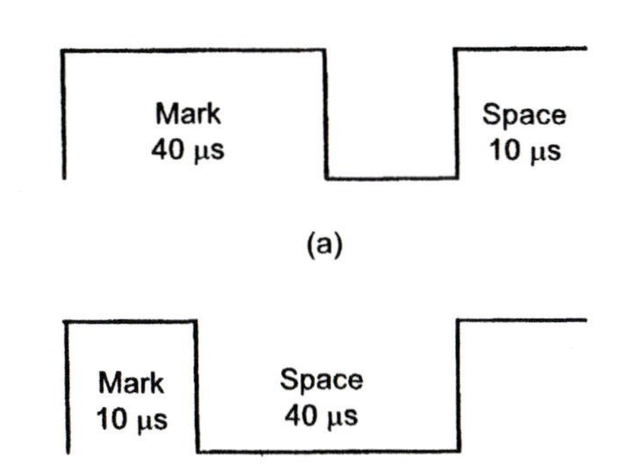

Waveforms with ac and dc components

A waveform can be regarded as pure ac only if it has equal areas enclosed between the time axis and the positive and negative parts of the waveform as shown in Figure 13.13.

Figure 13.13

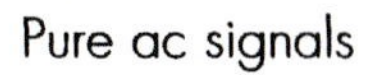
Pure ac signals

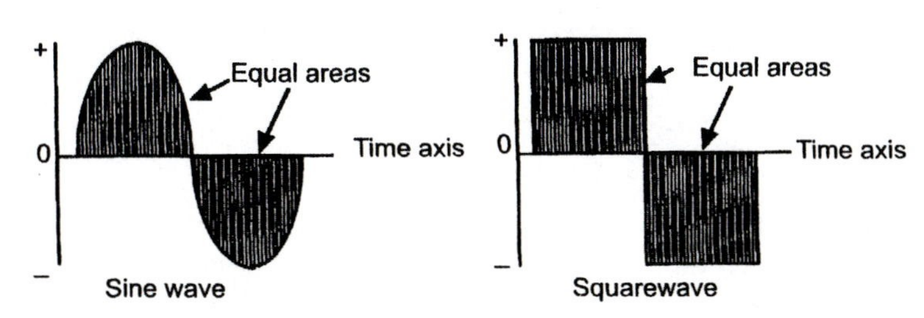

If equal areas do not exist on either side of the time axis the waveform can be considered as having a dc component. The dc component can be a negative or positive value but must be considered because sometimes it is very important to ensure that it remains unchanged, and in other cases it is important to remove it. AC waveforms with a dc component are shown in Figure 13.14.

Signal shaping

It has already been explained how removing harmonics from a person's voice changes how they sound. The same can happen to an

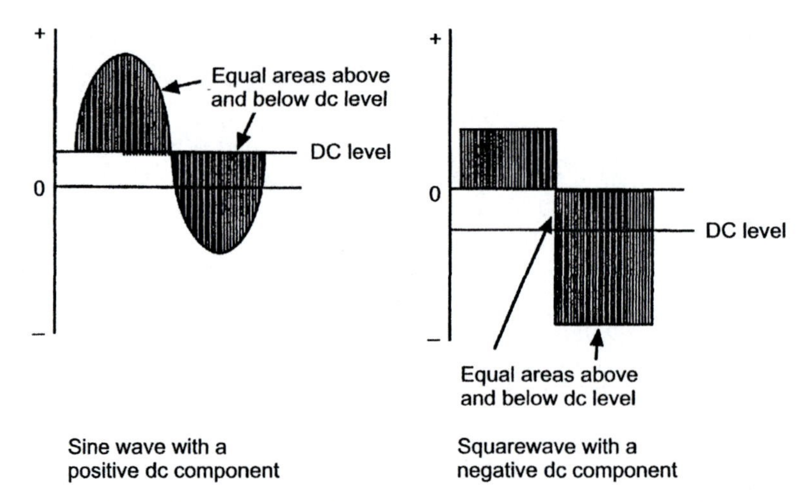

Figure 13.14 AC signals with a dc component

electrical signal as it passes through a system, but changing the shape of a waveform is sometimes done deliberately to achieve the desired outcome. Two commonly found circuits are used as signal shapers and are referred to as the *integrator* and the *differentiator.* They are also known as a *low pass filter* and a *high pass filter* respectively and are shown in Figure 13.15.

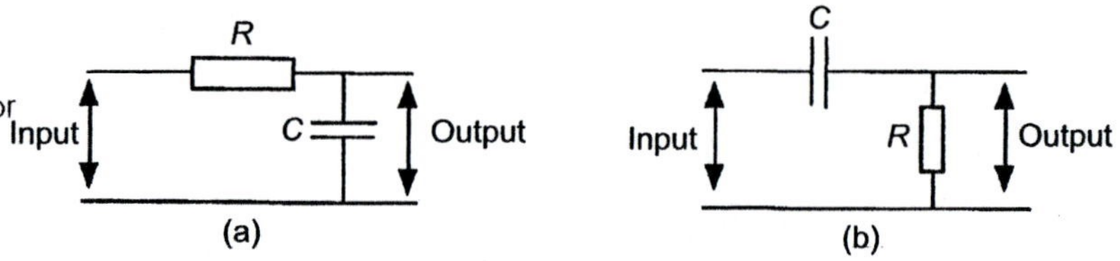

Figure 13.15 (a) Integrator. (b) Differentiator

R, C, Input, Output (a); C, R, Input, Output (b)

An integrator will change the shape of the input waveform by reducing the amount of high frequency harmonics or the high frequency content of a signal. This can be demonstrated by connecting a squarewave signal to the input of an integrator circuit and observing the output with the aid of an oscilloscope. The result of this experiment is shown in Figure 13.16(a) while the effect of a differentiator on the same input signal is shown in Figure 13.16(b). The exact effect will depend upon the values chosen for R and C as well as the frequency of the input squarewave signal. Since this is not a difficult exercise to undertake it is well worthwhile trying out a few different values and frequencies. It would perhaps be useful also to try an RC combination whose time constant is five times longer than the periodic time of the input frequency. The circuit's operation is easy to follow if some simple electrical principles are remembered. A capacitor has a reactance X_c measured in ohms sometimes called its ac resistance which reduces as the applied frequency increases so at low frequecies the integrator appears to be a potential divider network where the resistor's resistance is low compared to the reactance X_c of the capacitor. Only a small amount of the input voltage is dropped across the resistor with the rest of the voltage appearing across the reactance of the capacitor which gives a large output voltage at low frequencies. However, as the frequency increases the resistor's resistance remains constant but the reactance of the capacitor reduces so that the potential divider produces a low output voltage. Low frequencies pass but high frequencies do not, hence the title low pass filter. The differentiator relies on the same property of the capacitor except of course the circuit is configured differently so that it is the high frequencies that pass and the low ones that are filtered out. The squarewave is an ideal choice of input signal since it is a fundamental frequency plus its harmonics. The shape of the input signal is changed because the filter circuits remove some of the harmonics and change the amplitude of the others.

Figure 13.16
Integrated and differentiated waveforms

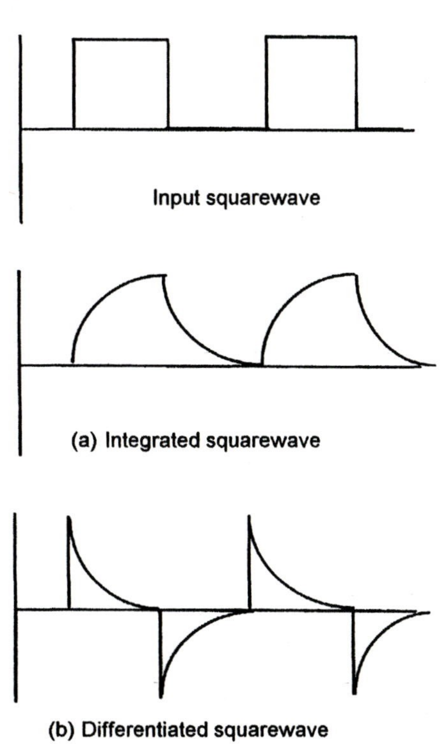

Measuring a signal amplitude

Some thought will already have been given to this subject when reading about the oscilloscope described earlier in the book, but there are several different methods and measuring instruments that can be used. It is important to know how a signal or voltage has been measured and the type of instrument used because different instruments produce different results. For example, the mains supply, which is generally said these days to be 230 volts although the majority of technicians will still say 240 volts, and at a frequency of 50 Hz, could be measured by using a simple ac voltmeter on the ac volts range. It is important to note that while it is easy to measure this voltage care should be taken and safety precautions should be observed, i.e. properly designed probes. Since it is an ac signal at 50 Hz an oscilloscope could also be used to measure its amplitude but this will not produce a reading of 240 volts. What then is the relationship between the readings obtained using different instruments? In the case of a dc voltage the value of the voltage is always constant but in the case of a sine wave it is more difficult because its value is always changing. There are three methods used to state the value of an alternating sinusoidal voltage or current.

Peak or maximum value

The peak value is the largest level the voltage or current reaches in either the positive or the negative direction. It is an important measurement particularly when considering the type of diode to use in a circuit, a point illustrated earlier in Chapter 10 when referring to the P.I.V. of a diode. This method of measurement is also useful to determine the peak to peak value of a signal and is shown in Figure 13.17.

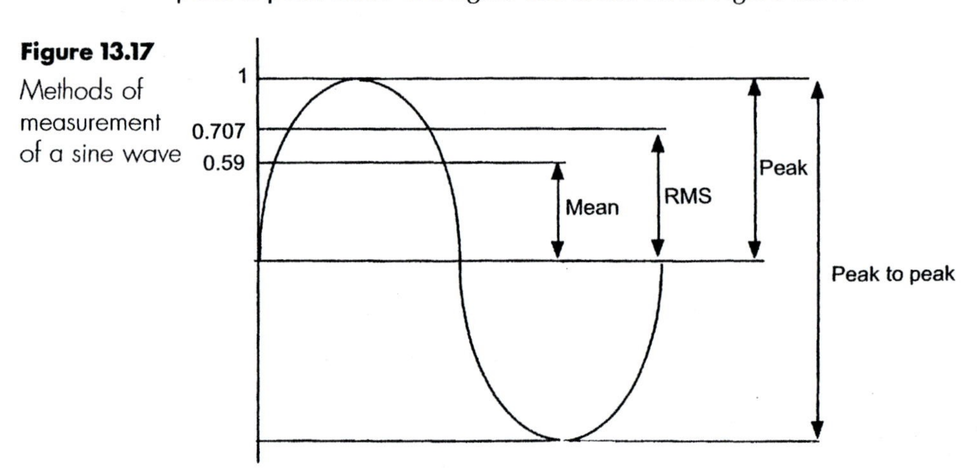

Figure 13.17 Methods of measurement of a sine wave

Mean or average value

This is the average value of the voltage or current that occurs during a half cycle. It may be found by determining the voltage or current at several points throughout the half cycle and calculating the average values of these instantaneous voltages. It is much less than the peak value and is generally only used when considering rectifier circuits but some moving coil meters give their readout as an average value. The relationship between mean/average value and peak value is shown in Figure 13.17.

RMS value

RMS or the *root mean square* value is the most commonly used method of measurement. There is a complex mathematical method of determining the RMS value of a voltage or current but for most service engineers it is simply the value of ac voltage that produces the same heating effect as a value of dc voltage. Hence the value quoted for the mains supply, 240 volts ac 50 Hz, will boil a pint of water in a 1 kW kettle in exactly the same time as it takes a 240 volt dc supply. Again it is much less than the peak value but is greater than the average value which can be seen in Figure 13.17.

Using the value quoted for the mains supply of 240 volts it is possible using the appropriate factor to determine either the peak value and hence the peak to peak value or the average value:

$$\text{Peak value} = \frac{\text{RMS value}}{0.707} = \frac{240}{0.707} = 339.5 \text{ volts}$$

$$\text{Peak to peak value} = \text{Peak value} \times 2 = 339.5 \times 2 = 679 \text{ volts}$$

It is perhaps now becoming more obvious why the mains supply is so dangerous and hurts so much when you receive a shock. It reaches amplitudes considerably larger than 240 volts. Using a capacitor as a filter across a mains switch as an example it should now be clear why it needs to have such a high working voltage, typically 1000 volts, and why on circuit diagrams the triangular safety symbol is placed alongside certain components, especially those which are subject to high voltage like the mains supply.

$$\text{Average value} = \frac{\text{RMS}}{1.11} = \frac{240}{1.11} = 216 \text{ volts}$$

Where has the 1.11 come from? There are several factors which are related that need to remembered. If the RMS value is taken as 1 volt the peak voltage is 1.414 times greater, so 1.414 is the *peak factor.* The *mean factor* is $\frac{1}{1.11}$ or 0.9:

$$\text{Average value} = \text{RMS} \times \text{Average/mean factor} = 240 \times 0.9 = 216 \text{ volts}$$

Power supply unit (PSU)

Having identified that an electronic system has a number of devices connected together and that the mains voltage is an ac signal, one of the first systems that is commonly encountered is a *power supply unit* or *PSU*. The block diagram is shown in Figure 13.18. It is of course constructed from components already identified elsewhere but now they are connected together to do something quite specific.

Figure 13.18
Block diagram power supply unit

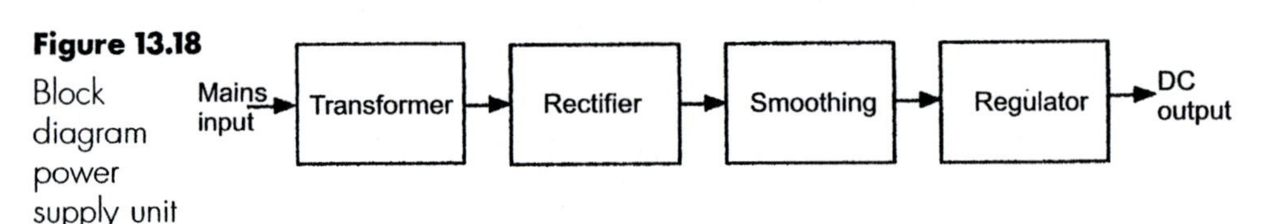

The purpose of the transformer in most modern pieces of equipment is to adjust the level of the mains voltage by stepping it down since the operating voltage of most equipment is considerably lower than the mains. It may at the same time provide inversion if a twin diode full wave rectifier is used and it always provides isolation if it is a true transformer and not an autotransformer which is very important for some systems. For example, a security alarm system must have a transformer. This ensures that it is almost impossible under a fault condition for the 240 volt mains to be fed to the peripheral parts of the system, such as passive infrared detectors or the cabling to them which is not designed to provide insulation at such a high voltage and even more importantly to the external sounder where an unsuspecting engineer standing on a pair of aluminium ladders in the rain might receive a fatal electric shock.

The diodes connected to the secondary of the transformer change the alternating output of the transformer into a series of unidirectional pulses. If a full wave rectifier is used there will be two pulses of the same polarity for every cycle of input but it is possible that only a half wave rectifier is employed in which case only one pulse of output per cycle will be obtained. These pulses would of course be rising and falling in the same manner as the mains supply and unsuitable for operating most equipment. What is required is a constant voltage like dc. The smoothing section has two parts: a reservoir capacitor, which stores a relatively large charge, and usually an electrolytic capacitor, which is connected to smooth the rectifier output pulses; these are followed by a filter, low pass type, to remove any ripple. The voltage should now be as close to dc as can be practically achieved.

The regulator is added to prevent the voltage and current output of the power supply varying beyond the design limits. If the load on the supply changes, for example the volume on an amplifier is turned to maximum, the PSU has to deliver more current because the amplifier is working harder. What would tend to happen without the regulator is that the voltage output would drop and the ripple voltage on the supply would increase. This would produce an audible humming in the loudspeaker and distortion because the bias levels to the transistors has changed. The regulator keeps the supply voltage constant throughout the accepted range of current it is designed to supply. Some regulators are also able to provide protection to the circuit since they are able to shut down if a fault occurs where the current drain exceeds the maximum set limit.

Amplifiers

There are many different types of amplifier. A simple amplifier has already been described when explaining the operation of a transistor. Different types of amplifier are used to perform specific functions and it is this that decides what type of amplifier is used. It is not possible to construct an amplifier that will amplify every frequency throughout the spectrum so an amplifier that operates over the range 20 Hz–20 kHz would be an audio amplifier. In the case of a radio receiver an amplifier that only amplifies a single frequency is required. This is a tuned or selective amplifier. The arrangement and type of components used in this amplifier circuit are different. In a selective amplifier the collector load resistor is replaced by a parallel tuned circuit which has a high impedance at the tuned or resonant frequency. This makes certain that the gain of the amplifier is maximum at the tuned frequency. The circuit diagram of a tuned amplifier and its frequency response are shown in Figure 13.19.

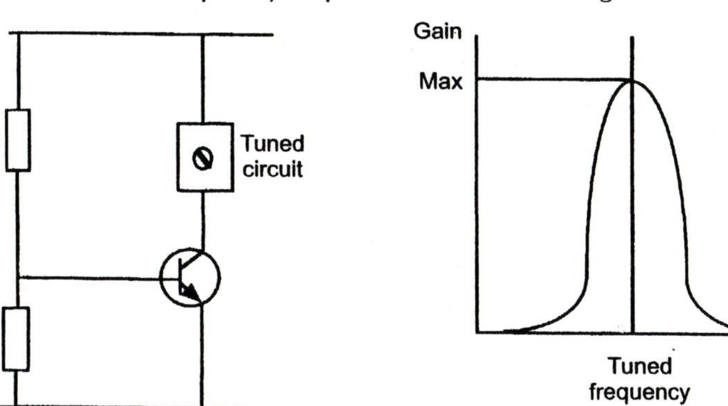

Figure 13.19 Tuned amplifier and frequency response

The tuned circuit may be formed by a transformer with a tuning capacitor connected across the primary winding. The transformer is designed to operate at high frequency so it is small and uses ferrite dust as its core. The core has a small indentation which is intended as an adjustment slot so that the tune frequency can be adjusted. A special trimming tool is needed for this operation not a screwdriver. To prevent radiation the assembly is mounted inside a small metal shielding can. This amplifier has a very narrow bandwidth which could be less than the 20 kHz of an audio amplifier.

An amplifier may be required with a much wider bandwidth, such as a vertical amplifier in an oscilloscope, and is known as a wideband amplifier.

There are three forms of distortion that amplifiers generally introduce. These are amplitude, frequency and phase distortion. If an amplifier has a frequency response characteristic which is not the correct shape over the required frequency range then it introduces frequency distortion. Typically this might be an audio cassette tape player that does not reproduce low frequencies so the reproduction sounds tinny. This is a form of frequency distortion. The tone controls, treble and bass, allow the user to adjust the frequency response to their own taste which helps to correct the effect.

Amplitude distortion is difficult to hide. The output signal has taken on a completely different shape when compared to the input signal. This occurs for a great variety of reasons but can be that the signal applied at the input is too large for the amplifier to handle and as a result *clipping* occurs. Two forms of clipping are show in Figure 13.20.

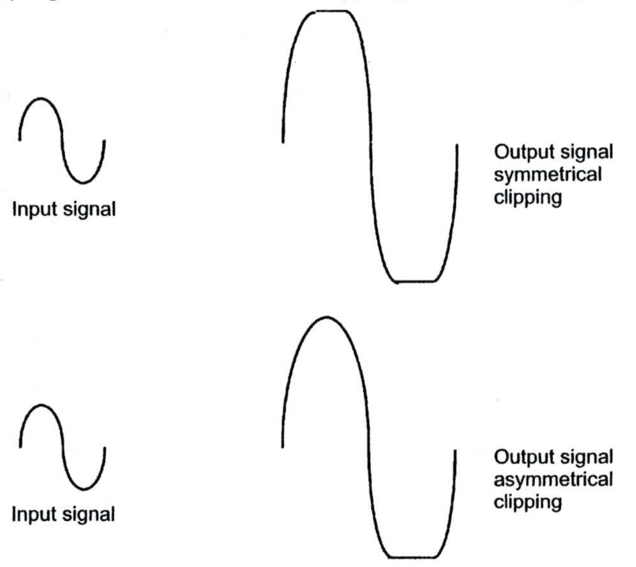

Figure 13.20
Effects of clipping in an amplifier

Phase distortion is caused by a change in signal phase as it passes through the amplifier. For audio systems it is not really a problem since it is a narrow band of frequencies, but in wideband systems, such as video amplifiers, a change of phase could result in a change of colour on the display which would be very noticable.

Limiting and clipping circuits

To prevent the input to a particular stage of an amplifier being overdriven, making its input signal too large, a limiting circuit can be used. This circuit employs a biasing arrangement which varies with the signal input and produces a relatively constant output and might be used as a form of automatic gain control to present a constant amplitude signal to the next stage in a device. Those readers who go on to study radio and television will meet this type of operation and circuit in nearly every receiver. The basic concept of a limiter or clipper circuit is shown in Figure 13.21 where the peak to peak amplitude of a sine wave is limited so that it is amplitude clipped and appears to be almost a squarewave.

Figure 13.21 Action of a clipping circuit

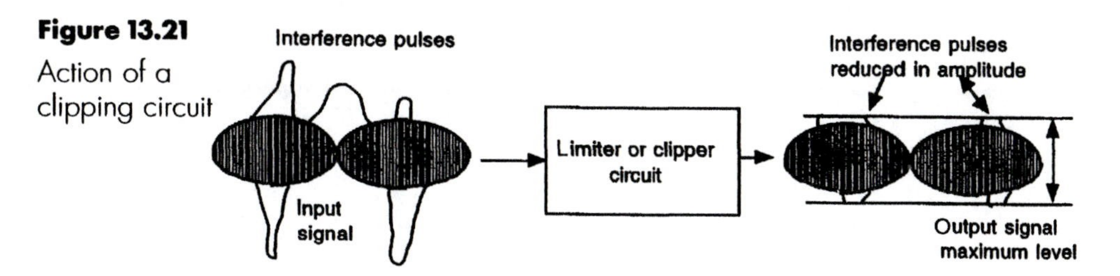

Clamping circuits

Earlier in this chapter it was stated that a signal might been made up of an alternating component and a dc component. When a signal is passed from one amplifier to another it is most likely that a simple coupling capacitor is used. Of course having discovered earlier, in the chapter on capacitors, that a capacitor will allow an ac signal to pass and block dc, the problem is now evident. If a capacitor is used for coupling when both the ac and dc components are required the dc component will have been removed. A clamping circuit ensures that both the ac and dc components are present at the output but with reference to the set level. This level could be a preset level in a control system or the black level of a video signal in a television receiver ensuring, for example, that the signal always was a positive value and started at 0 volts. This is shown as a simple example in Figure 13.22(a) and (b) where a signal with a dc component is amplified by two amplifiers which use ac coupling. The effect of the coupling capacitor is to remove the dc component but the ac signal is amplified. To re-establish the reference level, in this example 0 volts, the signal must now be passed through a clamping circuit where the dc component is restored. This effect is shown in Figure 13.22(c) and (d).

Speed control system

The stages of a speed control system are shown in Figure 13.23. This is a closed loop system. The essential feature of this system is that once

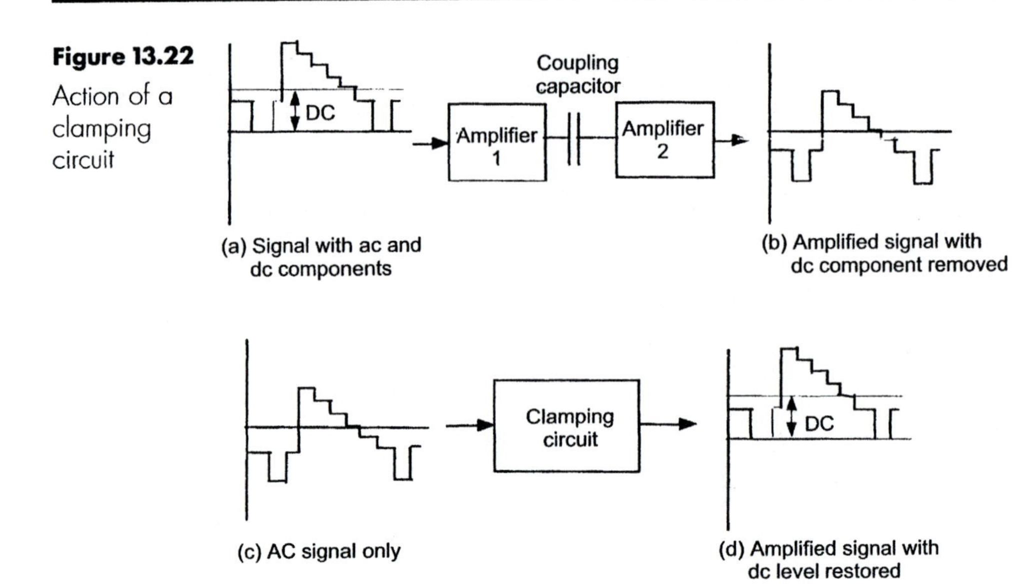

Figure 13.22 Action of a clamping circuit

the desired speed of an electric motor has been set its speed can be monitored and adjusted automatically.

The dc motor has its voltage and current provided by the dc power amplifier stage. The motor rotates and performs its drive function. To monitor the speed of the motor a dc generator is mechanically connected to the motor so that it turns at the same speed as the motor. The speed at which the motor turns determines the speed at which the generator turns so that the dc output voltage of the generator is directly related to the speed of the motor. If the speed of rotation of the motor increases the dc output voltage of the generator increases, and if the motor speed reduces the dc output voltage of the generator reduces. The generator's dc output is fed back to one input of a differential amplifier which has its second input connected to the set speed control, usually a variable resistor. The input voltage set by this variable resistor, once adjusted, is fixed but the input from the dc generator may vary. For example, if the motor speed increases the dc generator produces a larger amount of dc voltage. This causes one input to the differential amplifier to

Figure 13.23 Motor speed control system

Set speed control → Differential amplifier → DC power amplifier → DC motor — DC generator change so its output voltage falls. This fall in voltage appears at the input of the dc power amplifier and as a result its output voltage falls. With less voltage now coupled to the motor its speed reduces back to its set speed. Similarly if the motor speed fell the circuit would function in exactly the opposite manner to speed up the motor.

Basic computer system

A block diagram showing the essential parts of a computer system is shown in Figure 13.24. Most people think of a computer system as a desk based PC, but that is only one example. A video tape recorder, television receiver or a washing machine may all be examples of a system controlled by a small dedicated computer system. The essential parts and features are the same. The electronic components used in the system are called the *hardware* while the set of instructions that allow it to carry out certain tasks are termed *software*.

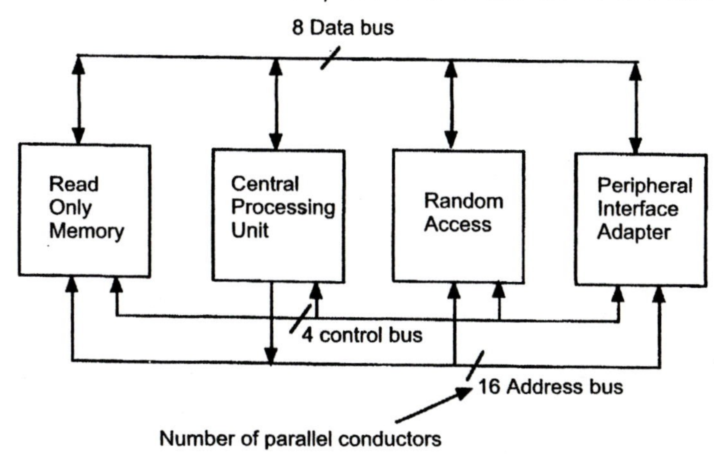

Figure 13.24 Simple computer system

The *central processing unit* (*CPU*) is the brain of the system. It contains a vast number of logic circuits in order for it to be able to decode instructions, generate address codes and control the operation of all the other parts of the system. In order for the CPU to know where to begin its task it needs an ordered set of instructions. This set of instructions must always be the same and is unique to the system. This is called a *monitor* program and given the term *firmware*. Firmware programs are stored in the Read Only Memory (ROM). The ROM is a non-volatile and non-destructive memory. Its contents are programmed in and even if the supply voltage is removed its contents will not be lost and its programs can be used over and over again without the need to reprogram the device. Many ROMs cannot be reprogrammed at all but it is possible to obtain EPROMS that behave in the same manner as a normal ROM but with the bonus that they can be reprogrammed firstly by erasing their contents using exposure to ultraviolet light and then reprogramming their data with the aid of a special machine. To perform the many tasks required the CPU needs to be able to store some data whilst it carries other functions, so another type of memory, the Random Access Memory (RAM), is used. This is a volatile memory whose contents will be lost if the supply is interrupted. The size of this memory will vary to suit the complexity of the system. A desktop PC may have many millions of storage cells measured in either megabytes or megabits, while a small dedicated control system may have only a few kilobytes. In order to carry out its functions the CPU must be connected to a source of input signals and output signals. This is the purpose of the interface circuits which may be called *ports* or *PIAs, peripheral interface adapters.* They act as buffers and control the flow of data to and from the external circuitry. They are quite complex components, as is a microprocessor system for which only a very simple description has been provided.

The microprocessor has three separate signalling networks known as *buses.* Each bus is a group of parallel conductors, the number being determined by the size and complexity of the system and the CPU used. A small dedicated control system might have only an 8 bit processor so its *data bus* will have only eight conductors. In order to communicate with its RAM it might therefore have 16 conductors to form its *address bus* and as few as four conductors to form its *control bus.* The data bus is bidirectional. Signals can pass in two directions along it both from the CPU to other devices and then back to the CPU from the RAM, etc. Both the address bus and the control bus are unidirectional and signals pass from the CPU to other devices only. The last component part of a microprocessor system is an oscillator. Usually it is a crystal oscillator so that the frequency is stable and does not drift. This allows the CPU to control timing functions. In order to keep the block diagram simple it is not shown.

Digital clock circuit

Reference has been made in the chapter on logic circuits to counters and decoders. A digital clock is made up of a suitable number of counters designed to count seconds, minutes and hours. The block diagram is shown in Figure 13.25. It would be useful to remember some basic principles at this point. Frequency is the number of complete cycles in a second, so if a signal at a frequency of 10 Hz is divided by a factor of 10 the result will be a signal with a frequency of 1 Hz. It also means where a signal started with

10 cycles per second the result of division is a signal with one cycle per second. It is possible to use this frequency to count seconds which is exactly the principle of the digital clock. The oscillator used initially is a high frequency oscillator and will be crystal controlled. This has the advantage of producing a very stable frequency free from drift because the frequency of oscillation is determined by the physical size of a piece of quartz crystal which is small and does not change. It is a sine wave oscillator but using a clipping circuit it is a simple operation to convert it into a square pulse for use with digital devices.

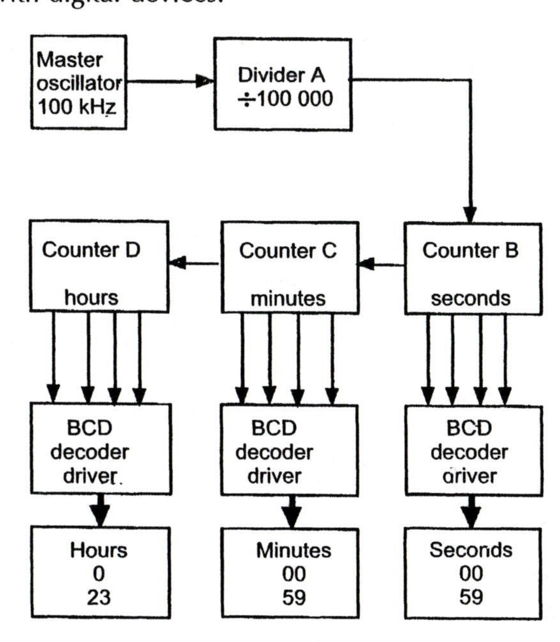

Figure 13.25
Block diagram of a digital clock

The oscillator might have a frequency of 100 kHz. To produce a pulse per second, that is one pulse out for every 100 000 in, a divider or counter with a dividing factor of 100 000 must be used. That seems quite a task but it is not difficult to produce a counter using integrated circuits which will do just that. Suppose a counter that divided by 100 is used first of all. For every 100 000 pulses at the input there would be only 1000 at the output. If another divider of 100 is used in series the 1000 pulses now at its input would be only 10 at this counters output. All that is now required is a further division by 10, the condition used in the original example of changing 10 Hz to 1 Hz. What appeared to be quite difficult is actually quite easy. Having established a pulse per second from divider A, a seconds counter B is used. B counts 60 pulses in and produces one pulse out since there are 60 seconds in a minute. A minutes counter C follows and again counts in the same way as B, 60 pulses in and one pulse out because there are 60 minutes in an hour. This is followed by an hour counter D which is a little different since its function is eventually to indicate the hours. It is therefore require to count from 0 to 23. The compltete counter must cover the range of 0.00.00 midnight to 23.59.59 when it resets to 0.00.00, its starting condition.

The three counters B, C and D produce binary coded outputs. Each of these outputs is fed to a BCD decoder driver which decodes the binary input signal and converts it into driving signals for seven segment displays.

Modulators and demodulators

The terms AM and FM when applied to radio receivers are quite familiar. The terms *amplitude modulation* and *frequency modulation* may not be quite as familiar but they are both processes that can be applied to a signal. In the case of a radio system it is to enable audio frequency signals to be transmitted many miles, but the process might also be used to carry control information to distant points. Modulation is a process where a characteristic of a sine wave is modified by a characteristic of another sine wave. If a signal undergoes the process of amplitude modulation to return it back to its normal condition it must be put through the opposite process which is amplitude demodulation. The same thing applies to a signal that is frequency modulated. Figure 13.26 shows an example of an amplitude modulated signal.

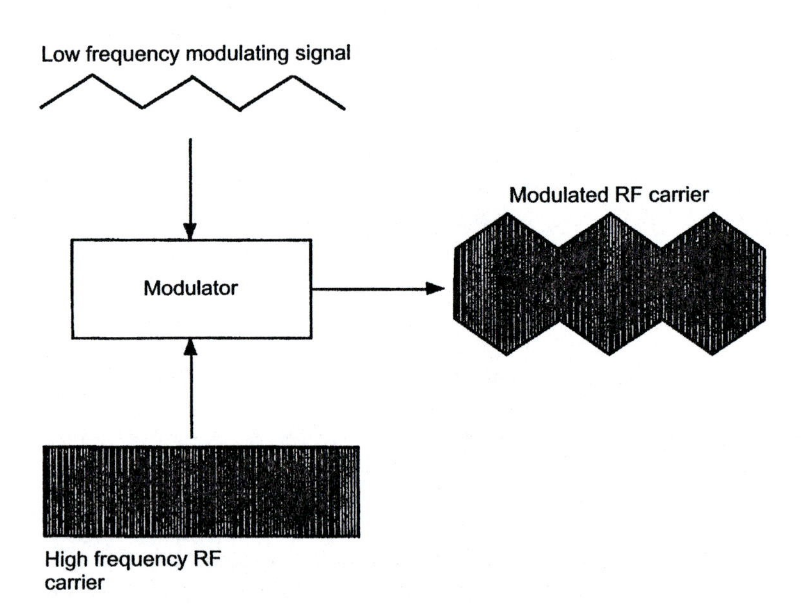

Figure 13.26 Amplitude modulator

The AM carrier shown in Figure 13.26 is a very complex signal, much more complex than is revealed by the simple diagram shown, but at this level it is only establishing the ideas and recognising names and processes that are important. To recover the original modulating signal a demodulator is used. This circuit responds to changes in amplitude with the result that it is able to reproduce the low frequency modulating signal at its output as in Figure 13.27.

Figure 13.27
AM demodulator

Demodulator
Output signal
low frequency
modulating signal
Input signal
modulated
RF carrier

Radio transmitters and receivers

A radio transmitter block diagram is shown in Figure 13.28. Many of the important stages and their functions have been identified already. Like most pieces of electronic equipment a power supply is required to provide each stage with the necessary voltage and current it requires to function. In some cases this could be a battery.

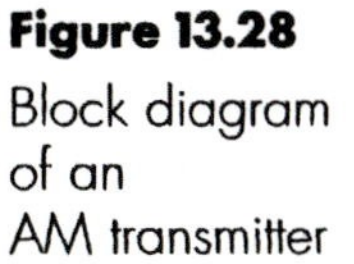

Figure 13.28
Block diagram of an AM transmitter

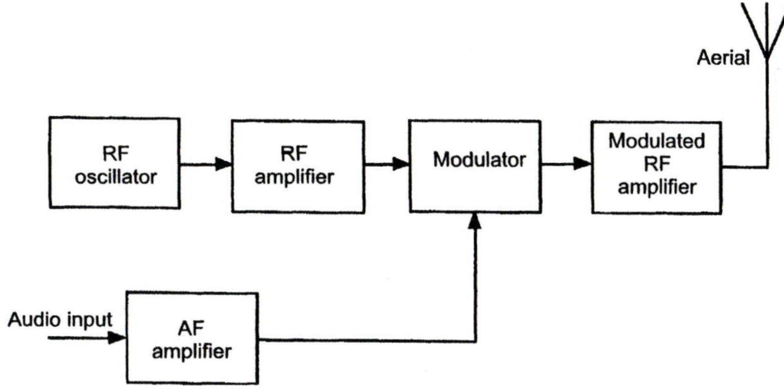

The first section is the RF oscillator. Its function is to generate a constant amplitude and constant frequency sine wave at the allocated frequency. Oscillators do not generally produce a power output and if a load is placed on them there can be a tendency to change frequency which is not desirable. As a result the output of the oscillator is passed to an amplifier tuned to the operating frequency of the oscillator to produce a signal at the desired amplitude to pass to the modulator. It also acts as a buffer between the oscillator and modulator stages preventing the oscillator circuit having its frequency changed. The audio frequencies from a microphone or tape recorder are amplified in an audio amplifier so that they have sufficient amplitude to operate the modulator but their amplitude is always carefully controlled since it must not become too large and cause overmodulation. The two signals react together in the modulator and an AM carrier is produced. It is no longer a single radio frequency as produced by the RF oscillator but a range of RF signals around the original oscillator frequency. This is referred to as a carrier and its sidebands. The output of the modulator is then passed to another RF amplifier tuned to the chosen frequency to produce enough energy to feed the aerial which radiates the signal.

The process of frequency modulation is different to that of amplitude modulation. In frequency modulation the RF carrier has its frequency changed by the amplitude of the modulating signal. This change in frequency is called the *deviation* and for radio broadcasts is given a maximum deviation of +/− 75 kHz. The blocks are rearranged as in Figure 13.29.

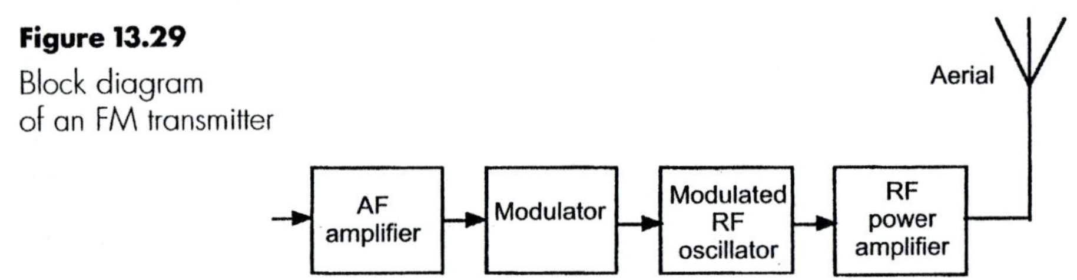

Figure 13.29
Block diagram of an FM transmitter

The transmitter again relies on having an RF oscillator to generate the desired carrier frequency. However, in this arrangement the audio signal is amplified and then used to change the frequency of the oscillator so that the deviation is controlled by the amplitude of the audio modulating signal. The greater the amplitude of the audio modulating signal the greater the change in frequency of the oscillator. The amplitude of the oscillator output signal should be constant. This is a great advantage since the power amplifier which drives the aerial with the RF signal can have a smaller power handling capacity. Furthermore as most types of interference produce amplitude variations on a signal this system is relatively interference free which is an advantage of an FM system.

Having built systems to transmit both AM and FM signals the receivers for these systems need to be examined. There are a variety of systems which can be constructed for use on either of these two systems. The most common arrangement uses the *supersonic heterodyne principle* commonly called a *superhet.* It has a very simple fundamental principle which is that, irrespective of the RF frequency a user tunes to, the incoming RF signal will be changed to a common frequency before amplification. This new frequency is called the *intermediate frequency* and abbreviated to the term IF. Before examining the superhet receiver it is worth looking at the simple design of a *tuned radio frequency* receiver or *TRF*. This type of receiver can be used for either AM or FM reception provided the appropriate type of demodulator is used. A basic block diagram is shown in Figure 13.30.

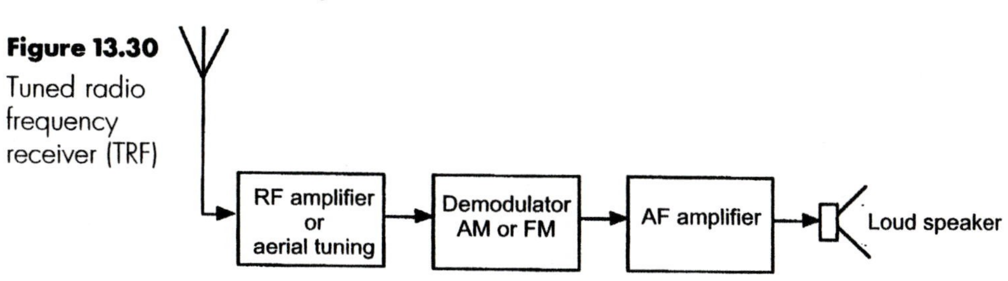

Figure 13.30 Tuned radio frequency receiver (TRF)

The transmitted signal, electromagnetic radiation, is received by the aerial. To separate the wanted signal from all the signals passing through the aerial some form of tuning may be used. A tuned aerial or a tuned amplifier or a combination of both selects the wanted frequency. The signal needs to be amplified since it is only very small at the aerial before it is passed to the demodulator where the original modulating information is recovered. The signal received was of course at RF, whilst the output of the demodulator is AF. Again this AF signal is only small and while it may be large enough to drive a set of headphones it will not be large enough to drive a loudspeaker, so an audio amplifier is used for this purpose. Of course this system could be even simpler, returning to the basic design of a high efficiency tuned aerial circuit coupled to a demodulator whose output is fed to a set of high impedance headphones. This is the crystal set, the forerunner of the modern day receiver. As more and more transmitters came into use and it became necessary to be able to tune to a large variety of transmitters the TRF design had a number of problems which could not be overcome easily at that time. This led to the development of the superhet mentioned earlier and this is shown in the block diagram of Figure 13.31.

The superhet design was used because at the time it could be made more selective than a TRF receiver. Its basic principle of changing the tuned incoming frequency into a different common frequency allowed a number of selective amplifiers to be used in series thereby producing a high gain and a narrow bandwidth characteristic, precisely what is required. Just as with the TRF receiver some method of selecting the wanted frequency from the many transmitted frequencies is required. In an AM receiver the usual method is to use a *preselector* stage. Normally this is a *ferrite rod* aerial fitted inside the case of the receiver. It is a tuned circuit. In the FM receiver a tuned RF amplifier is used because the frequencies used for FM transmissions are much higher and it is essential that the ratio of the wanted signal to the noise signal is maintained as large as possible. This incoming RF signal is then fed to a *mixer* stage. The receiver has its own in-built RF oscillator known as the *local oscillator* which is set to operate at a slightly higher frequency than the tuned incoming RF signal. The preselector stage and the local oscillator are linked together so that adjusting the tuning of the receiver automatically adjusts both stages by the same amount which maintains the difference in frequency between the stages. The local oscillator signal is also fed into the mixer stage. The mixer stage used to be known as the *frequency changer* which is perhaps more descriptive of its function since the output of this stage will be a new common frequency, the difference in frequency between the local oscillator and the incoming RF signal. For domestic receivers AM transmissions have an IF around 470 kHz whilst for FM transmissions the figure is 10.7 MHz. The IF signal is then amplified in a series of two or three tuned amplifiers before being passed to the demodulator where the original modulating signal is recovered, just as in the TRF receiver, before being amplified and passed to the loudspeaker. In the block diagram of Figure 13.31 typical frequencies for an AM radio receiver have been suggested.

Figure 13.31
Block diagram of an AM receiver

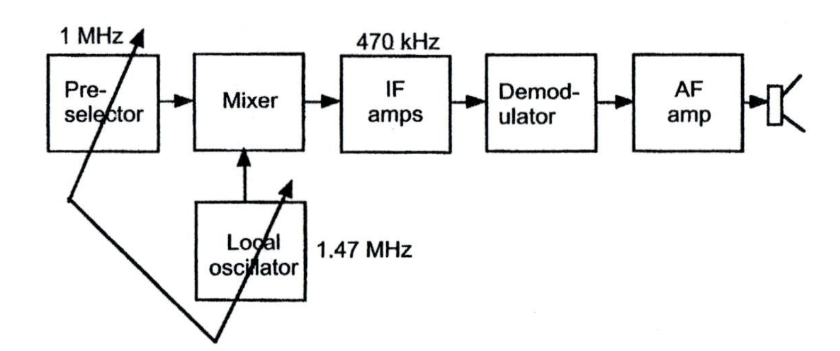

Television transmitters and receivers

The block diagram of Figure 13.32 shows a simplified version of a colour television transmitter. A television transmitter must transmit several sets of information at the same time. The obvious information being the picture and the sound but because of the complex system of transmission there must also be synchronising information for the time bases and for the colour decoding section of the receiver. The type of modulation used for the visual information is amplitude modulation with frequency modulation being used for the sound.

Figure 13.32
Block diagram of a television transmitter

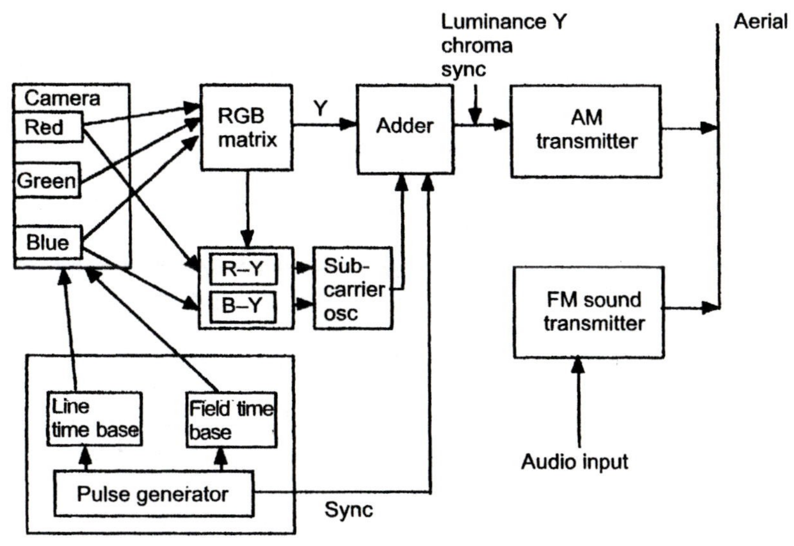

The camera generates three signals, one for each of the three primary colours used in the system: red, green and blue. These three signals are combined to form the luminance signal Y which represents the brightness information in a scene and is the signal that those watching in monochrome use to produce their picture image. Only two signals are required to convey the colour information, so the red and blue signals are used. They are combined in a matrix to form colour difference signals, R–Y and B–Y, which in turn quadrature amplitude modulate a colour subcarrier oscillator. The camera cannot produce a single electrical signal that represents a whole picture so, like the reader of a book, reads line by line from top to bottom. The two time bases cause the camera to scan through a picture; the line time base scanning from left to right and the field time base scanning from top to bottom. Since the receiver must scan at the same rate and in the same sequence these frequencies must be exact and information must be sent from the transmitter in the form of synchronising pulses to carry out this function. The pulse generator controls the time base operation. The three signals, luminance, chrominance and sync, are combined and then used to produce an amplitude modulated vision carrier signal. To be able to transmit pictures with good definition the luminance signal has a wide bandwidth and uses signals from a few hertz to 5.5 MHz. The sound signal is frequency modulated and uses its own FM transmitter linked to the video transmitter. The sound transmission frequency is always 6 MHz higher than the vision carrier frequency in the British UHF system.

The television receiver block diagram is shown in Figure 13.33. Again it is a much simplified version where a single block performs many quite complex functions. It is of course a superhet receiver and works on the same principle as the superhet radio described earlier where an incoming frequency is changed to an IF frequency. This function is carried out in the tuner stage. The output from the tuner is the vision IF signal and also the sound and chrominance signals which are all amplified in the one set of vision IF amplifiers. This keeps the receiver design relatively simple. At the vision demodulator the sound signal and the chrominance signal are separated and then processed in their own sections. The difference in frequency, 6 MHz, between the vision carrier and sound carrier is now exploited and used to create a 6 MHz inter-carrier sound IF that is frequency modulated. This FM signal is amplified and passed to an FM demodulator where the sound information is recovered, passed to an AF amplifier and to the loudspeaker.

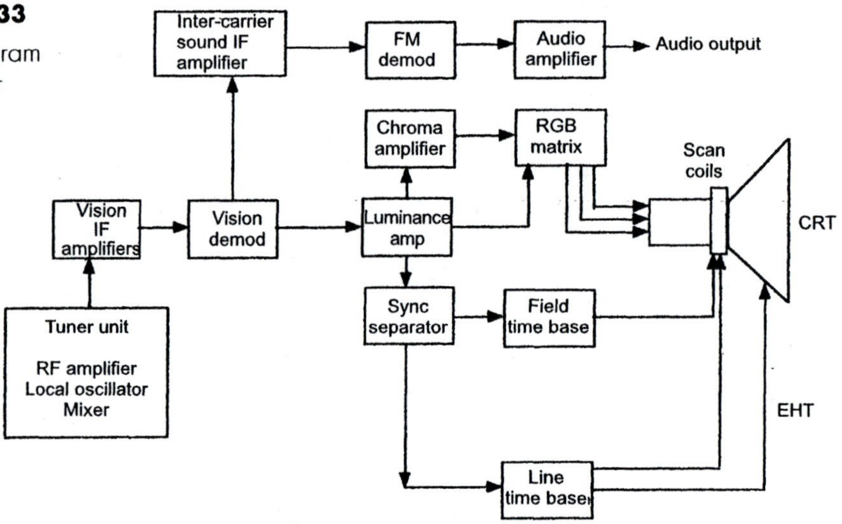

Figure 13.33 Block diagram of a colour television receiver

The luminance signal output from the vision demodulator also contains the chrominance and sync information. The chrominance signal is separated by a filter circuit tuned to the chrominance subcarrier frequency which in the British system is 4.43 MHz. This signal is amplified and demodulated before being recombined with the luminance signal in the RGB matrix and passed to the cathodes of the cathode ray tube.

The luminance signal is also passed to the input of the sync separator which removes the luminance information and leaves just a series of pulses at the line and field frequencies to synchronise the time bases. The two time bases in the receiver line and field are used to produce scanning currents to the field scan coils and line scan coils at exactly the same frequency and in the same phase as the time bases scanning the picture at the camera. The field scanning frequency is 50 Hz and the line scanning frequency is 15.625 kHz. The line time base is also used as a generator for the EHT potential and the auxiliary supply voltages that are required elsewhere in the receiver.

The audio tape recorder

This is again a useful system since it shows how some of the different devices can be connected to carry out another useful function but using a different type of transducer, the magnetic record/playback head. The block diagram of a simple audio tape recorder is shown in Figure 13.34.

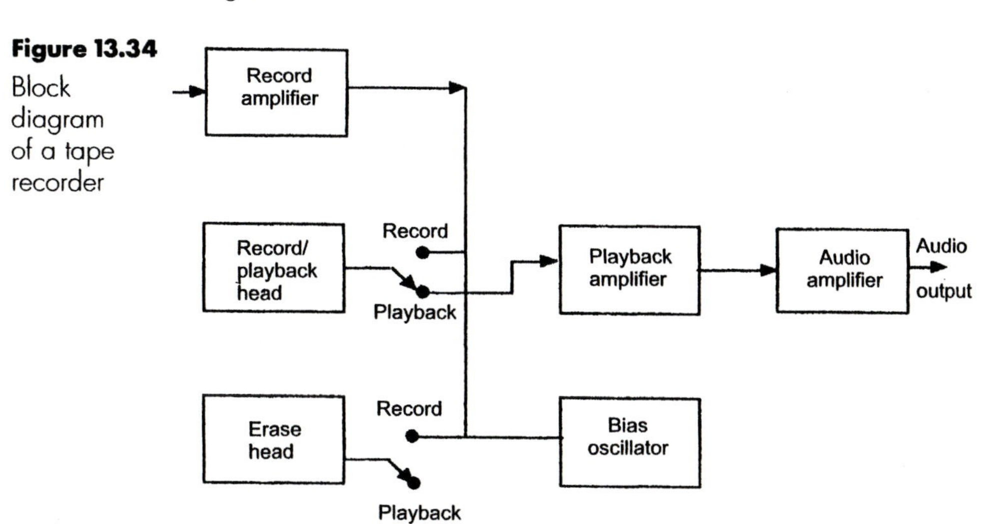

Figure 13.34 Block diagram of a tape recorder

When the machine is in playback mode it is simply an audio amplifier. The tape with its varying magnetic fields passes across the playback head and induces a voltage in its coils. This voltage is then amplified and passed to the loudspeaker. In the record mode some different stages are switched into operation. For a tape recorder of reasonable quality a bias oscillator is used. This oscillator generates a sine wave of about 70 kHz. When passed to the erase head it produces an alternating magnetic field which removes any other magnetic fields present on the recording tape and therefore anything previously recorded. It ensures that the tape is ready to be recorded on. The same bias signal is also coupled to the record head. This is needed because of the non-linear characteristics of magnetic tape. As well as this bias signal the audio input signal which has been amplified is coupled to the record head so that the magnetic tape records the fields produced by the audio input signal. In the playback mode the switches move to a different position and the bias oscillator has no function and is usually turned off altogether to prevent it producing interference problems.

There are many electronic devices that can be connected together to form electronic systems. In the repair of this equipment it is always useful to try and think in terms of blocks, what the function of each block is, where the input signal is coming from and where the output signal goes to. This helps to identify the processes a signal goes through and hence what it should look like, or the level it should be at and if these correspond to information provided in circuit manuals, etc. They are a valuable aid in locating the faulty stage in a system which is the first step necessary in locating the faulty component.

14

Logic circuits

Electronic circuits can be divided into two categories, analogue or digital. Analogue systems are also sometimes referred to as linear systems and are designed to amplify or process signals whose amplitude is smaller than the supply voltage. Logic or digital circuits as they are more commonly called have only two states:

(i) when the device is fully conducting and therefore has a very low voltage drop across it;
(ii) when the device is non-conducting and turned off completely.

The advantage of operating in either of these two conditions is that very little power is dissipated in the device because the current is small or the voltage drop is low.

Logic networks

In this type of circuit the signal can have only a *high* or *low* condition. The high state is indicated by using the figure 1 and the low state by using the figure 0, therefore the system is called a binary system, one which has only two states. This is fine but what does it mean from a practical electronic point of view? Initially the logic gate can be considered as an ordinary switch, for example a light switch. This has two conditions. One where its contacts are closed and the light is on and the other where the contacts are open circuit and the light is off. There is no half way point. Figure 14.1(a) shows a simple circuit to illustrate this point. The switch contacts are open circuit so no current flows and as a result the lamp does not light. Figure 14.1(b) shows the second condition where the switch contacts are closed. Current is flowing in this circuit so the lamp lights.

Figure 14.1
Simple switch circuit

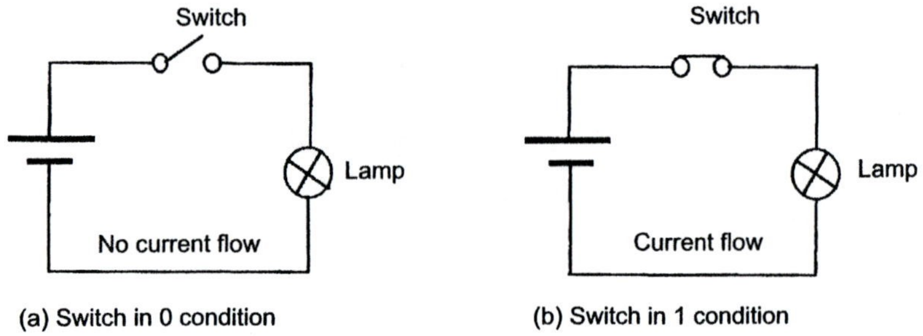

Switch
Lamp
No current flow
(a) Switch in 0 condition

Switch
Lamp
Current flow
(b) Switch in 1 condition

The AND function

The impression given of a logic gate is that it is a different type of switch, but this is only partially true and is a description used for convenience. A gate is designed to behave a little differently than a simple switch and performs in a distinct manner when a precise set of input conditions occurs. Examining Figure 14.2 shows a modified version of Figure 14.1. Two switches have been connected in this circuit in series with the lamp. For the lamp to light current must flow and this can only happen if both the switches A and B are closed. The condition of a switch can be shown using the binary method stated earlier, that is, a closed switch is signified by the number 1 and an open circuit switch by the number 0. It is possible in this circuit to have a number of different settings for switches A and B but the only condition that allows the lamp to light is as previously stated where A is at logic 1 *and* where B is at 1. The equivalent logic gate circuit is therefore called an *AND gate.* A circuit performing the AND function must conform to the statement that *all inputs* must be at the high state, 1, to give an *output* at the high state, 1. To show the behaviour of a logic gate a table called a truth table is used. Figure 14.3 shows the truth table of an AND gate. The table shows every possible combination of the switches A and B and the expected output usually designated as F, the function.

Figure 14.2
Series circuit

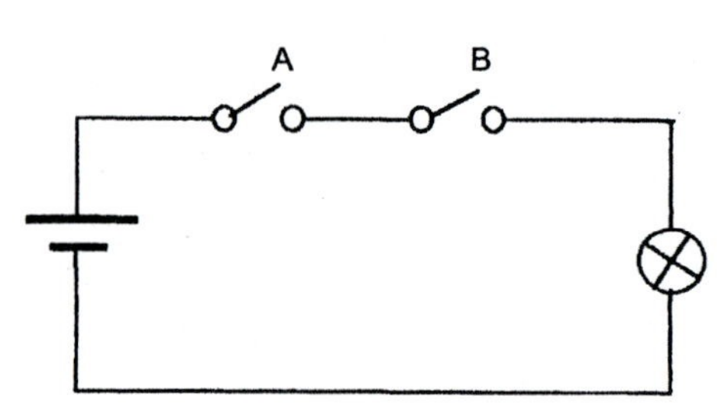

The truth table of Figure 14.3 shows, as expected, that the only occasion a high output occurs is when switches A and B are closed, A = 1, B = 1 then F = 1.

Figure 14.3
Truth table
AND gate

A	B	F
0	0	0
0	1	0
1	0	0
1	1	1

The OR function

Just as resistors or capacitors can be connected in series or parallel, so too can the switches. This is shown in Figure 14.4 and furthermore it should be obvious that this circuit does not behave in the same manner as the AND equivalent circuit of Figure 14.2.

Figure 14.4
Parallel circuit

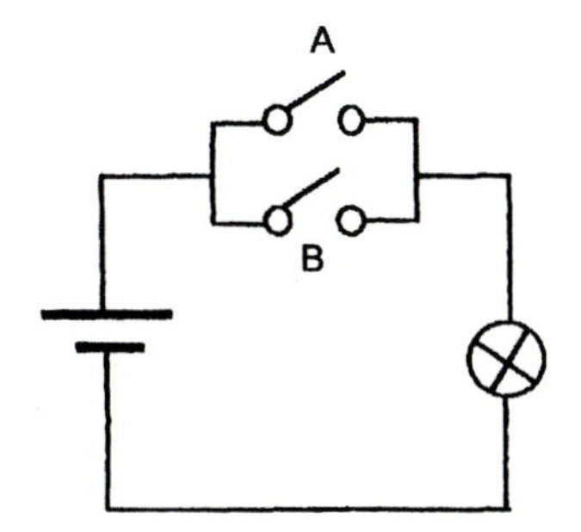

In Figure 14.4 when both switches are open circuit, at logic 0, no current flows so the lamp does not light, therefore the output is at logic 0. However, if switch A OR switch B OR both switches are closed, at logic 1, the lamp will light so the output will be 1. The circuit behaviour can be shown using a truth table. Figure 14.5 shows the truth table of an OR gate.

Figure 14.5
Truth table
OR gate

A	B	F
0	0	0
0	1	1
1	0	1
1	1	1

To show the type of logic device being used in a circuit each logic gate must have its own unique circuit symbol. A problem arises at this point since there are a number of symbols in common use. The two popular ones are the British Standard symbols and the American MILSPEC symbols. The symbols used for an AND gate and an OR gate are shown in Figure 14.6.

Figure 14.6
Common symbols for AND and OR gates

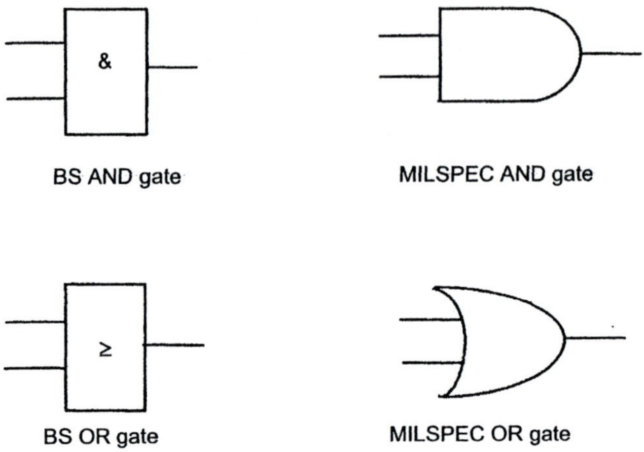

&

BS AND gate

MILSPEC AND gate

≥

BS OR gate

MILSPEC OR gate

If the simple lamp circuit used in Figure 14.2 is rearranged so that it now appears as that of Figure 14.7 a different set of circumstances has been created. When the switch is open circuit the lamp will light since current can flow through it but if the switch is closed the lamp is bypassed, shorted out and does not light. Using the same reasoning as with the previous circuits a truth table can be drawn and is shown in Figure 14.8, but note that for this type of gate there is only one input.

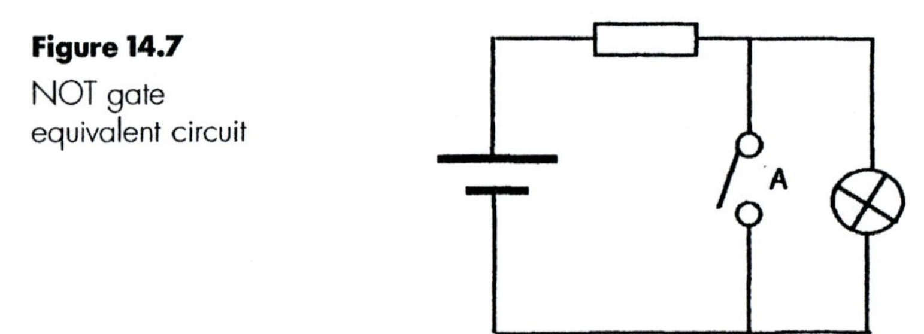

Figure 14.7
NOT gate equivalent circuit

The output F is always the opposite or *inverse* of the input. This logic function is called the NOT function. The symbols used for a NOT gate are shown in Figure 14.9. The small circle on the output connector signifies that inversion takes place inside this device.

All logic circuits are based on the three functions identified so far;

however, for convenience and during manufacture it is common for combinations of AND and NOT to be used or, alternatively, OR and NOT to be combined. These combinations produce two more logic functions of NAND and NOR.

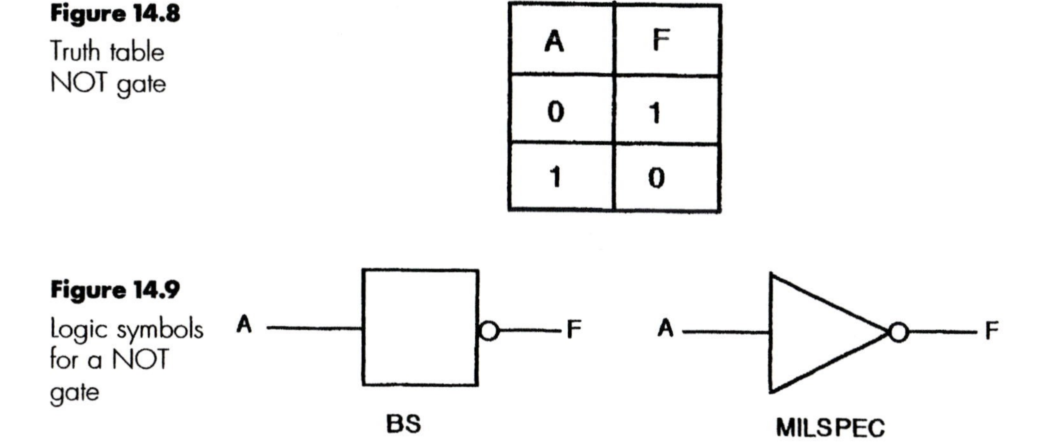

A	F
0	1
1	0

Figure 14.8
Truth table NOT gate

Figure 14.9
Logic symbols for a NOT gate

The NAND gate can be considered as an AND gate whose output has been passed through a NOT gate whilst the NOR gate can be considered as an OR gate whose output has been passed through a NOT gate. Each of these combinations has its own circuit symbol and truth table as shown in Figure 14.10(a) and (b).

It is essential to remember the truth table for each of these logic gates. In all of the examples shown , with the exception of the NOT gate, it is possible to have more than two inputs. For example, it is not uncommon to find a NAND gate with four or even eight inputs. Examining the truth table of the two input NAND gate (Figure 14.10(a)) reveals that the output only goes to 0 when both its inputs are at 1. This can be extended to cover all numbers of inputs by saying that the output of a NAND gate only goes to 0 if *all* inputs are at 1.

The exclusive OR function

Logic gates can be arranged to provide other logic functions and the exclusive OR is an example. It is not a single gate that can be shown using a simple equivalent circuit with switches and a lamp as used earlier when discussing other gates. It is a network of other gates connected together to provide a specific action; however, there is a logic symbol used to show this function. Unlike other gates which can have multiple inputs the exclusive OR and the exclusive NOR gate can only have two inputs and a single output.

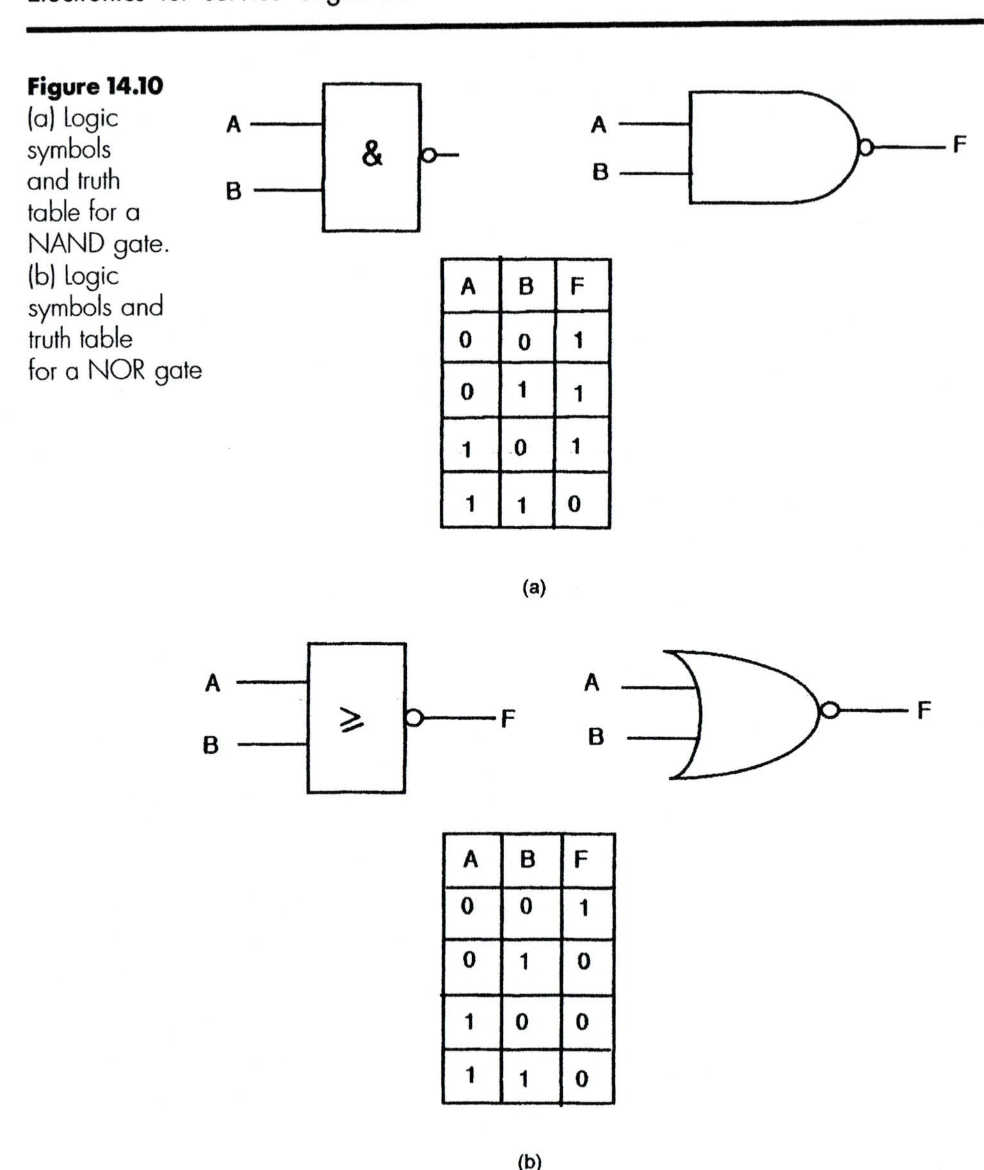

(a)

A	B	F
0	0	1
0	1	1
1	0	1
1	1	0

(b)

A	B	F
0	0	1
0	1	0
1	0	0
1	1	0

Figure 14.10 (a) Logic symbols and truth table for a NAND gate. (b) Logic symbols and truth table for a NOR gate

Unlike transistors, where it is possible to obtain a single device, logic gates are produced inside integrated circuits. It is usual to find a number of similar gates inside the same single IC, typically four NAND gates or four NOR gates. Figure 14.12 shows a typical device which contains four NAND gates. Each gate has two inputs, one output and shares the same low voltage supply terminals. It is worth noting how the connecting pins are numbered, starting at the top left-hand corner with the key at the top then counting down the

Figure 14.11
Circuit symbols and truth tables, exclusive OR and exclusive NOR gates

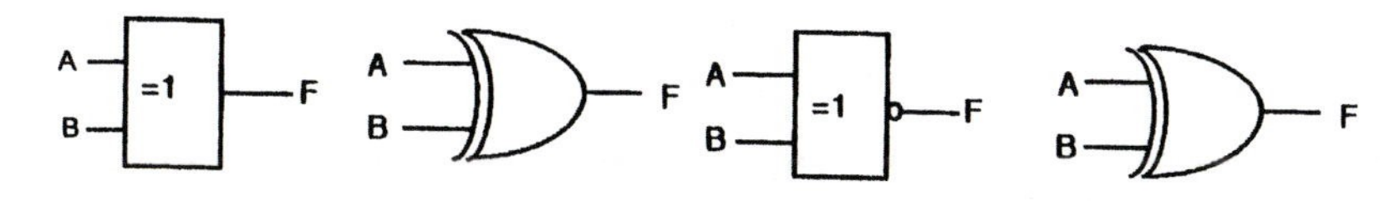

A	B	F
0	0	0
0	1	1
1	0	1
1	1	0

A	B	F
0	0	1
0	1	0
1	0	0
1	1	1

Figure 14.12
7400 quadruple 2 input NAND gate

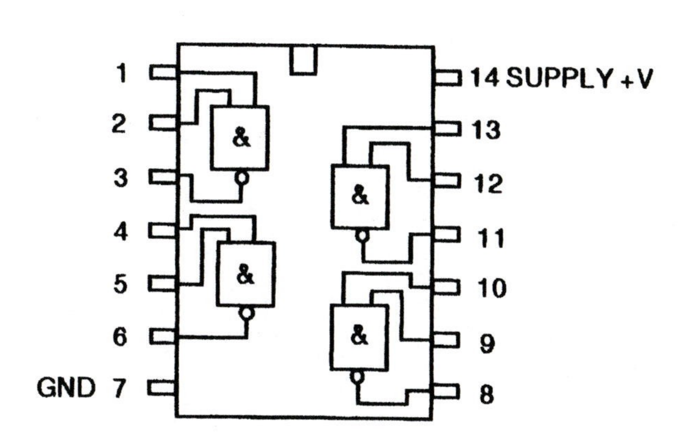

left-hand side and back up the right-hand side. Gates can also be manufactured using different technologies: one is based on bipolar transistor technology and known as transistor transistor logic or TTL, while another type is based on field effect transistor technology and is known as CMOS, complementary metal-oxide semiconductor. Although the logic function might be the same it is not possible to interchange devices from these two families because their characteristics are different.

The two main families are 74 series TTL and 4000 series CMOS. Each family has many different packages, far too many to list here but a full listing is available in the form of data sheets from manufacturers or suppliers, for example, RS produce a very comprehensive list. The main features of each family are listed in Table 14.1.

Table 14.1 TTL characteristics (standard 74 series*)

TTL characteristics	
Absolute maximum supply voltage	7 V dc
Recommended operating voltage	5 V dc
Maximum input signal in positive direction	5.5 V dc
Maximum input signal in negative direction	1.5 V dc
Minimum logic one level	2.0 V dc
Maximum logic 0 level	0.8 V dc
Fan in	1
Fan out	10
CMOS characteristics (4000 series)	
Operating supply voltage	3 to 18 V dc
Logic one level	0.7 V dc to supply voltage
Logic zero level	0.3 V dc to 0 V dc

* There are a number of variations of the 74 series family, for example 74LS series. Its operating characteristics are slightly different so it is important to establish precisely which particular variant is used to be certain the circuit will operate correctly and reliably.

Fan in

Logic gates are connected together to form more complex circuits. When interconnecting gates it is important to make sure that the logic gates being connected are compatible otherwise a malfunction will occur. The fan in of a gate has to be considered. Although the current consumption of a logic gate is small it will require some current at its input to make it work. Rather than quote a current, manufacturers use the figure fan in. It represents the current drive requirement of a single gate input. For a standard 74 series TTL gate this figure is stated as being 1 unit.

Fan out

When connected together to form logic circuits a single logic gate may be called upon to provide the input to a number of other logic gates simultaneously. To be certain a logic gate will perform reliably its fan out is given and this value represents in simple terms the current handling capability of the gate. For a standard 74 series TTL gate the fan out is 10. From a service engineer's point of view it is important to replace a faulty gate with a gate from the same logic family and type but for a constructor building a circuit a little more care will need to be taken to be certain the circuit operates as expected.

Sequential and combinational logic

Just as with resistors and capacitors which can be connected together to form series circuits or parallel circuits it is possible to do exactly the same with logic gates. Logic gates connected in series are said to be sequential circuits and when connected in parallel, combinational circuits. Examples of sequential circuits which are connected serially are counters and shift registers. They rely on the use of a circuit called a *bistable* or *flip-flop.* A flip-flop is a circuit which can have two possible output conditions. It can be made to flip into one of these conditions where it will remain until it is flopped back into its other condition. The simplest flip-flop can be constructed by using a pair of NAND gates or a pair of NOR gates and is called an R–S flip-flop or an R–S latch.

Figure 14.13
NAND gate R–S flip-flop

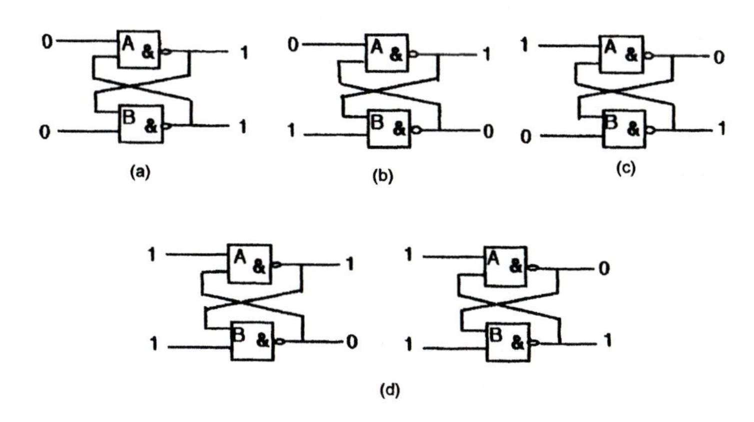

In Figure 14.13(a) to (d) a pair of NAND gates has been used to build such a circuit. It has two inputs and two outputs, four combinations of input logic but only three possible output conditions. Figure 14.13(a) to (c) shows three of the combinations of input signal and the resulting output states. In (a) both gate A and gate B have the controlled input at 0. NAND logic demands that a NAND gate with at least one of its inputs at 0 must produce a logic 1 at its output. Gate A must therefore produce a 1 at its output and so too must gate B. The flip-flop is said to be in a *forced* condition, an unnatural condition for a flip-flop and a condition to avoid, the reason for which will be demonstrated a little later on.

In Figure 14.13(b) only gate A has logic 0 at its input. Its output therefore is a logic 1 and this is coupled to one input of gate B along with a logic 1 from the control input. Gate B must by the rules of NAND logic produce a logic 0 output. Moving onto Figure 14.13(c), and again applying NAND logic, shows that gate A output must produce a logic 0 and gate B a logic 1. Finally the fourth input condition of Figure 14.13(d) must be considered. It is a little more difficult to determine the output on this occasion since both gate A and gate B have a logic 1 at their control input. The conclusion reached is that the flip-flop could adopt either one of its two output conditions and is sometimes referred to as a *don't know* state. Redrawing and labelling the NAND gate circuit of Figure 14.13 shows the R–S flip-flop in its SET and RESET conditions, the only two conditions wanted. This is shown in Figure 14.14.

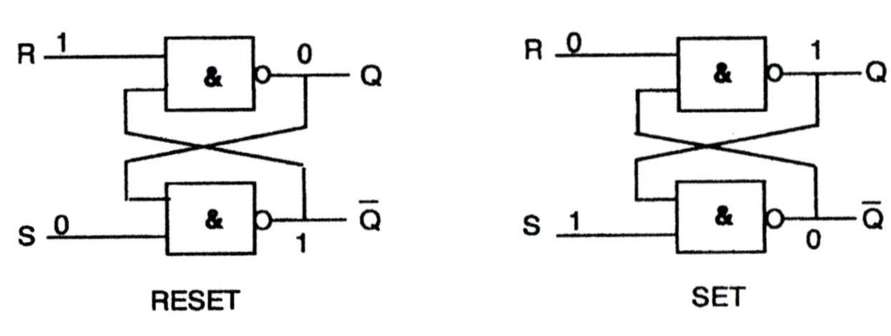

Figure 14.14 R–S flip-flop

The operation of this circuit is easy to predict as a one-off using NAND logic, but when used a part of a logic system dealing with a stream of binary ones and zeros it is perhaps a little more difficult to follow. It is important to remember the output conditions for SET and RESET since this will help. A sequence of data at the flip-flop control inputs could be as shown in Table 14.2.

Table 14.2

R=0 S=1 the flip-flop goes to RESET	Q=0 $\bar{Q}$=1
R=1 S=1 the flip-flop stays in RESET	Q=0 $\bar{Q}$=1
R=1 S=0 the flip-flop goes to SET	Q=1 $\bar{Q}$=0
R=1 S=1 the flip-flop stays in SET	Q=1 $\bar{Q}$=0
R=0 S=1 the flip-flop goes to RESET	Q=0 $\bar{Q}$=1
R=0 S=0 the flip-flop goes to forced condition	Q=1 $\bar{Q}$=1 FORCED
R=1 S=1 the flip-flop goes to either	Q=1 $\bar{Q}$=0*
	Q=1 $\bar{Q}$=1*

* Since it is impossible to predict which of these two conditions the flip-flop will go to should the R=0 S=0 input condition be followed by the R=1 S=1 INPUT condition the FORCED condition must be avoided. This problem may also be referred to as RACE HAZARD since one gate is trying to switch fully before the other.

R–S flip-flop truth table

R	S	Q	$\bar{Q}$	
0	0	1	1	FORCED
1	0	0	1	RESET
0	1	1	0	SET
1	1	1 0	0 1	

Further variations of the R–S flip-flop are manufactured. In some applications it is important that precise timing is required. It is necessary to allow some time for the control inputs at R and S to become steady or at their required levels but delay the setting or resetting of the output until the precise moment a change is desired. This is achieved by adding two more NAND gates to the standard R–S flip-flop already discussed so that it forms a *gated R–S flip-flop,* which is shown in Figure 14.15.

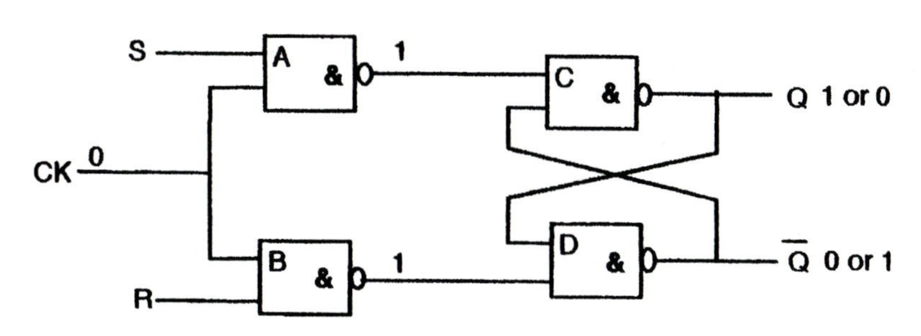

Figure 14.15 Gated or clocked R–S flip-flop

This logic gate circuit has two distinct parts. Gates A and B form a gating network while gates C and D form a flip-flop. Examining the logic diagram of Figure 14.15 shows that if the CK input is held at logic 0 the output from gate A and the output from gate B must be a logic 1. Whatever level, either 1 or 0, is applied at the R and S inputs cannot change the output condition of gates A and B since these outputs are held at a logic 1 level because of the absence of a clock signal. The flip-flop can remain in either the SET or RESET condition with the control inputs to gates C and D at a 1 level. Only when the CK input is taken to the logic 1 level can the R and S inputs have any effect. Figure 14.16 shows the conditions that can occur.

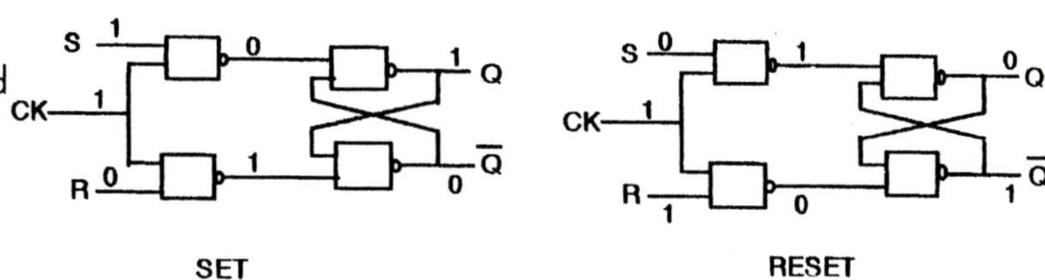

Figure 14.16 Gated/clocked R–S flip-flop in SET and RESET conditions

To summarise the actions of a gated or clocked R–S flip-flop the following statements can be used.

If S = 1 and R = 0 when a clock pulse is applied the flip-flop goes to SET.
If R = 1 and S = 0 when a clock pulse is applied the flip-flop goes to RESET.
If R=0 and S=0 when a clock pulse is applied *no change* occurs.

When using an R–S flip-flop it is more convenient to use simple logic symbols rather than to draw gate networks. Figure 14.17 shows the symbols used for the R–S and gated R–S flip-flops.

Figure 14.17
Circuit symbols for R–S and gated R–S flip-flops

Once again it is possible with this type of flip-flop to have the problem of race hazard if both the R and S inputs are at 1 when a clock pulse is applied. This condition would cause the flip-flop to go to the forced condition with both the outputs at logic 1. When the clock pulse is removed the outputs of the gating section of the device return to the 1 condition but the flip-flop is left to decide whether it adopts the SET or RESET condition. Clearly this is unacceptable since the output state cannot be predicted. This indeterminate condition can be avoided by further modification. All that is required is to ensure that the R and S inputs can never be the same, the inclusion of the NOT gate shown in Figure 14.18. prevents the problem. The flip-flop is now called a *D type flip-flop.* Just as in the examples of the R–S and gated R–S flip-flop the gate action could be followed through step by step but this is not necessary. What is necessary is simply to remember that the Q output of a D type flip-flop will go to the state of the D input when a clock pulse is applied. Figure 14.19 shows the logic symbols used for D type flip-flops.

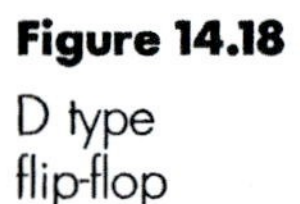
Figure 14.18
D type flip-flop

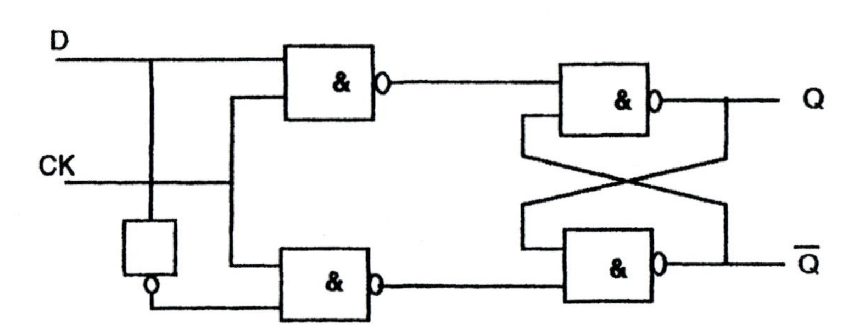

There are many variations of flip-flops used. A careful examination of the circuit symbol will reveal which type is in use. In Figure 14.19(a) there is an arrow on the clock input which signifies that the device is triggered on the positive edge of the clock pulse, whilst in Figure 14.19(b) the arrow and the circle show it is triggered on the negative edge of the clock pulse. Examples of the different flip-flops can be found on the data sheets suggested earlier on. For example, a

Figure 14.19
Logic symbols for D type flip-flops

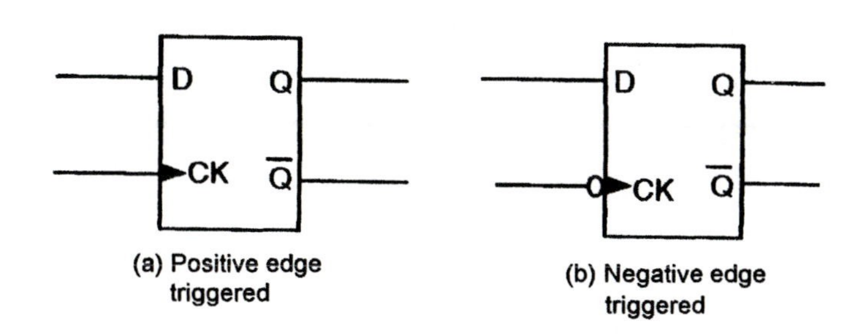

74LS74 is a dual D type edge triggered flip-flop and a 4042B is a quad D latch. Another very popular flip-flop is the J–K flip-flop. It is not necessary to examine the logic gate array inside this device since this is usually a complex arrangement of two gated R–S flip-flops with feedback from output to input operated in a master slave arrangement. What is important is to remember the circuit symbol and the information in the truth table for use later on when the device is used in a counter or shift register. The basic J–K flip-flop symbol and truth table are shown in Figure 14.20.

Figure 14.20
Logic symbol and truth table for a J–K flip-flop

Logic symbol J–K flip-flop

J	K	Qn +1	Qn +1	
0	0	Qn	$\overline{Qn}$	No change
0	1	0	1	RESET
1	0	1	0	SET
1	1	$\overline{Qn}$	Qn	REVERSAL OF STATE

Qn is the condition of the flip-flop before a clock pulse is applied

Qn + 1 is the condition of the flip-flop after one clock pulse has been applied

Before moving on to examine some of the various circuits using flip-flops it will be helpful to consider the variety of devices available, their logic symbols and the implications for their operation in a circuit.

Figure 14.21
Logic symbols for D type and J–K flip-flops

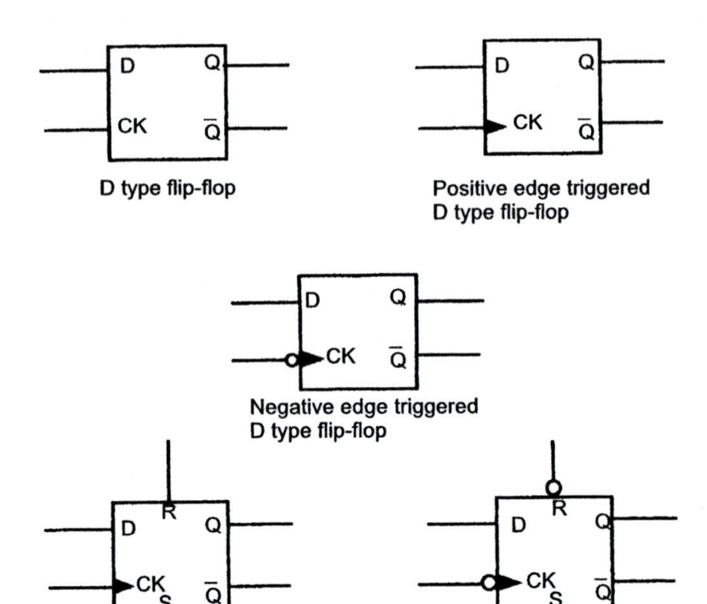

D type flip-flop

Positive edge triggered D type flip-flop

Negative edge triggered D type flip-flop

Positive edge triggered D type flip-flop with active high direct set and reset inputs

Negative edge triggered D type flip-flop with active low direct set and reset inputs

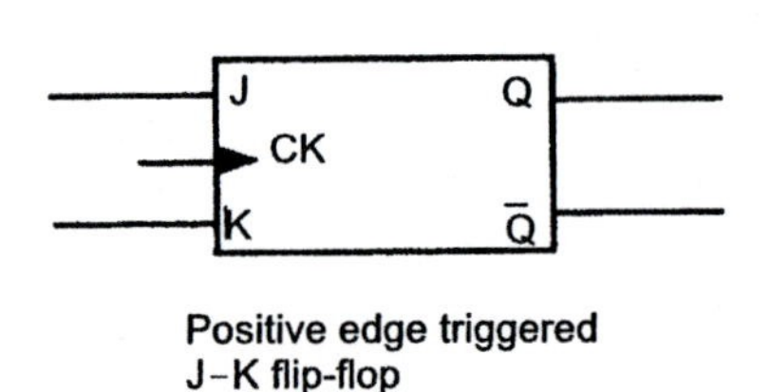

Positive edge triggered J–K flip-flop

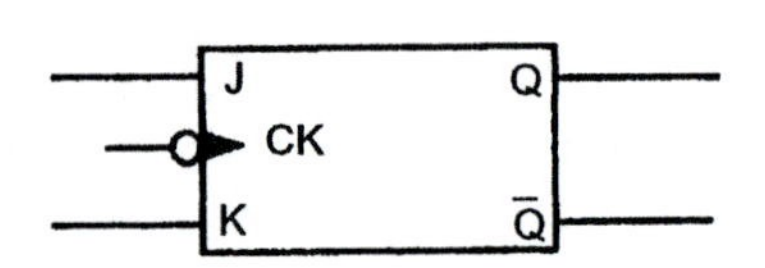

Negative edge triggered J–K flip-flop

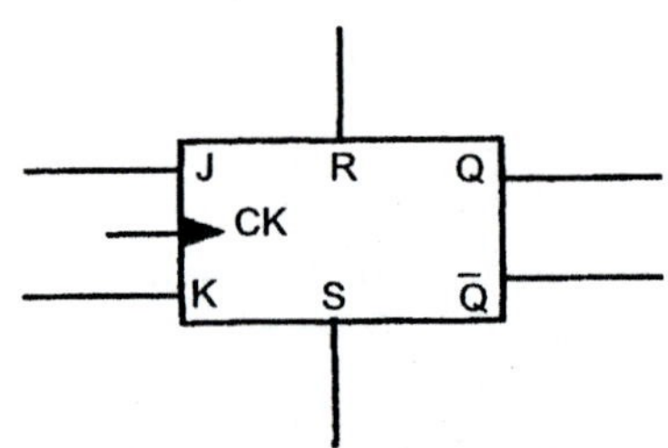

Positive edge triggered J–K flip-flop with active high direct set and reset inputs

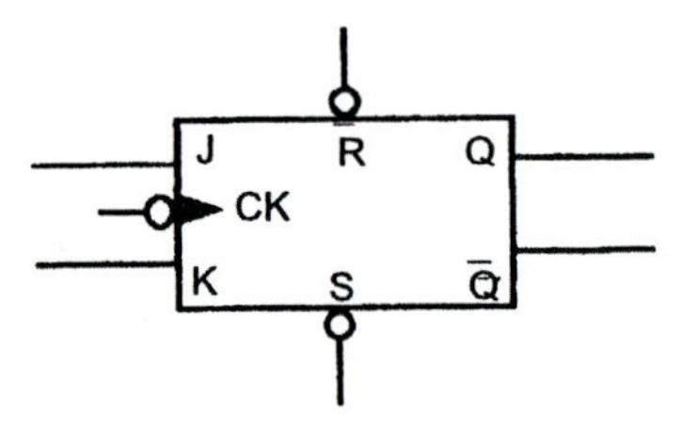

Negative edge triggered J–K flip-flop with active low direct set and reset inputs

There are many different ways in which logic circuits can be constructed to achieve a similar output. It should be made clear, for example, that a device may be required to produce one output pulse for every ten pulses applied at the input, a decade counter, but although the output may be just one pulse out for every ten put in the point at which the pulse occurs and its shape may also be important.

Binary counters

These devices convert a pulse count or clock input into a binary coded output. If, for example, six pulses are applied at the input the output would be four binary levels giving the code 0110. The first example shown in Figure 14.22 uses J–K flip-flops connected to operate asynchronously or in a manner sometimes described as a ripple counter.

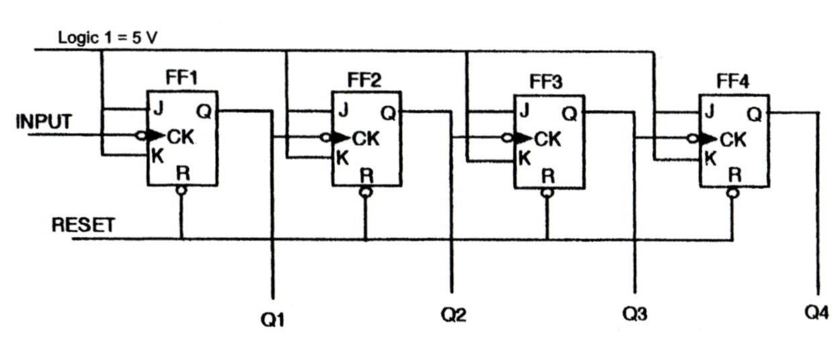

Figure 14.22 Asynchronous binary counter

Each of the flip-flops has its J and K inputs coupled together and connected to a logic 1 level, for a TTL device this would be +5 V. The first action required is to reset the counter so that all the outputs Q1 to Q4 start at the same level and for a counter this logically is 0. To achieve this the direct reset line is held low, 0 V, for a brief period of time. While the direct reset line is low applying clock pulses has no effect since the direct reset (or direct set) inputs override all other inputs. After resetting, the counter is ready to start. Each time a clock pulse is applied the J–K flip-flop will change state since both inputs are held at logic 1. The type of J–K used is a negative edge triggered device so any change at a Q output can only occur on a trailing edge of a pulse. Figure 14.23 shows a typical clock pulse.

Starting from the reset condition the action of the counter in Figure 14.22 will be as follows. On receiving the first clock pulse FF1 changes state. Its output Q1 goes to logic 1. This change in level, however, does not act as a clock pulse for FF2 since it has only risen and produced a leading edge. FF2, FF3 and FF4 therefore remain unchanged. When a second clock pulse is applied FF1 again changes state so its output now returns to 0. This fall in output level acts as a trailing edge of a clock pulse to FF2 which changes state. Q2 goes to logic 1. FF3 has still not received a complete clock pulse, a trailing edge, so it remains unchanged as must FF4. The action continues and is best illustrated by examining a waveform or timing diagram as in Figure 14.24.

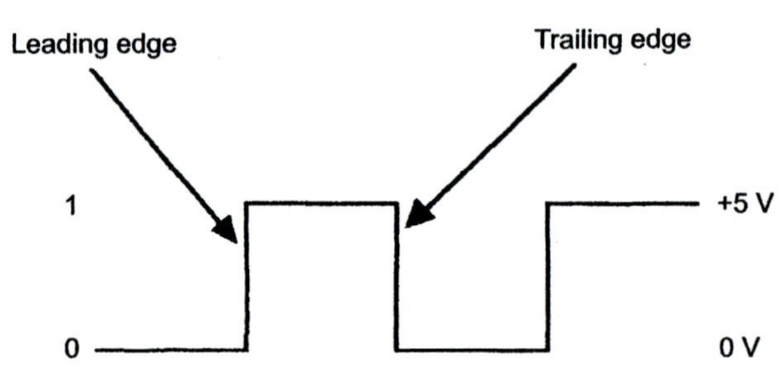

Figure 14.23 Typical clock pulse

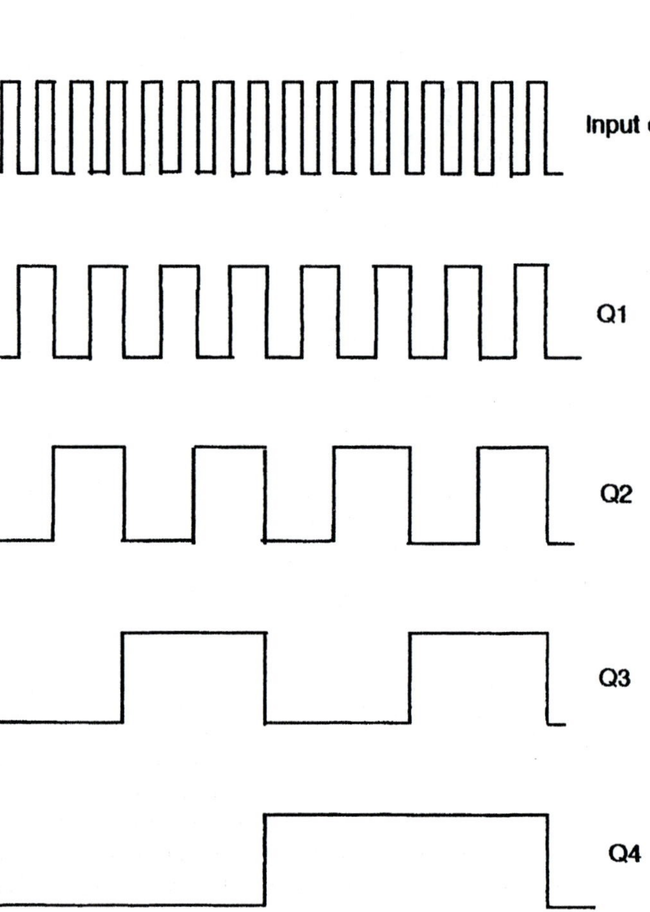

Figure 14.24 Output waveforms of a binary counter

This waveform diagram shows that each J–K flip-flop acts as a divide-by-two device. The input clock waveform is divided in FF1 by a factor of two and the output of FF1 is divided in FF2 by a factor of two and so on. Each flip-flop provides a division of its input waveform by a factor of two which is one of the fundamental principles of binary circuits. The waveform diagram of Figure 14.24 is referred to as the time domain but it is also possible with some test equipment to view the outputs in the *data domain*. This is a useful way to view the outputs since they are a reminder of a binary count sequence which is shown in Figure 14.25.

Figure 14.25 Data domain chart of a four stage binary counter

Input clock pulse number	FF1	FF2	FF3	FF4	
0	0	0	0	0	RESET
1	1	0	0	0	
2	0	1	0	0	
3	1	1	0	0	
4	0	0	1	0	
5	1	0	1	0	
6	0	1	1	0	
7	1	1	1	0	
8	0	0	0	1	
9	1	0	0	1	
10	0	1	0	1	
11	1	1	0	1	
12	0	0	1	1	
13	1	0	1	1	
14	0	1	1	1	
15	1	1	1	1	
16	0	0	0	0	RESET

This type of counter was referred to earlier as a ripple counter because the changes of state ripple through from one device to another. Not shown in the waveform diagram is the cumulative delay produced by each stage, FF2 must wait for FF1, FF3 must wait for FF2 and so on. The delay time of a device is the length of time it takes for the output to respond to the change applied at the input. This limits the maximum speed at which this type of counter can be operated. For example, if each flip-flop has a delay of 40 nanoseconds the cumulative delay will be the number of stages (4) multiplied by the delay time of each stage which gives a total delay of 160 nanoseconds. A waveform with a periodic time of 160 nanoseconds would have a frequency of 6.25 MHz and applying a clock input signal higher in frequency than this would produce an inaccurate result. To overcome this problem counters can be constructed so that all stages change state at the same time, thus avoiding the rippling effect and allowing use of higher frequencies. This type of counter is called a *synchronous counter* and is shown in Figure 14.26.

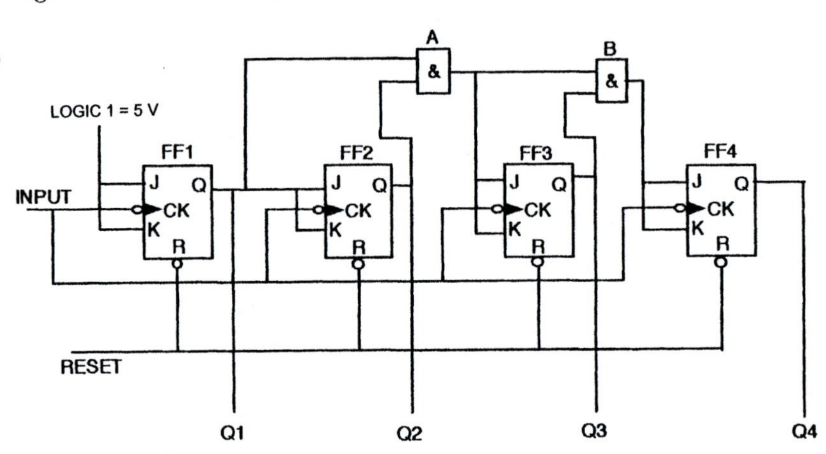

Figure 14.26 Synchronous binary counter

The synchronous counter shown in Figure 14.26 again makes use of negative edge triggered J–K flip-flops with active low direct reset inputs. The internal structure of this type of J-K flip-flop is important to understanding its operation. As stated earlier in the chapter it is quite complex but to keep things as simple as possible it can be said that each flip-flop consists of two stages. The one connected to the input is called the *master* while the one connected to the Q output can be called the *slave*. What is even more important is to remember that on the leading edge of the clock pulse the master is active, setting itself into the desired condition, with the Q or slave output fixed. On the trailing edge of the clock pulse the master is disabled and held constant while the slave takes up the state decided by the master. It is quite difficult to describe in operation because by its very nature all changes take place at the same time. The J and K inputs of each flip-flop are linked together so if they are at 1 when a clock pulse is applied the flip-flop changes state and if they are at 0 when a clock pulse is applied the output stays the same. Starting from the RESET state, all flip-flops with their Q outputs at 0, Q1 will go to 1 on the trailing edge of the first clock pulse but at the instant of the clock the J and K inputs to the other flip-flops were at 0 so they stay the same. On the application of a second pulse Q2 goes to 1 and Q1 returns to 0. On the third pulse Q1 changes again and goes to 1 but FF2 does not change because at the instant the clock pulse was applied its inputs were at 0. Q1 and Q2 are both at 1 which of course is correct for 3 in binary. The AND gate A now has a 1 on both its inputs and is providing a 1 for the input of FF3. When the fourth clock pulse arrives FF3 changes state, its output goes to 1 while Q1 and Q2 both return to 0. This operation will continue until after the fifteenth pulse has been applied where all the flip-flop outputs are at 1. The next pulse will return all the outputs back to 0 and the count will repeat itself again. The output waveform (Figure 14.24) or the data output chart (Figure 14.25) are the same for the synchronous counter. It should of course be obvious that a device that counts to sixteen is not practical for everyday use since the numbering system used for everyday transactions is the decimal or denary system. The counter can be modified so that it will count up to nine and then reset to zero. An example is shown in Figure 14.27 but it is only one method of many which could be used to achieve this.

Figure 14.27
Decade counter

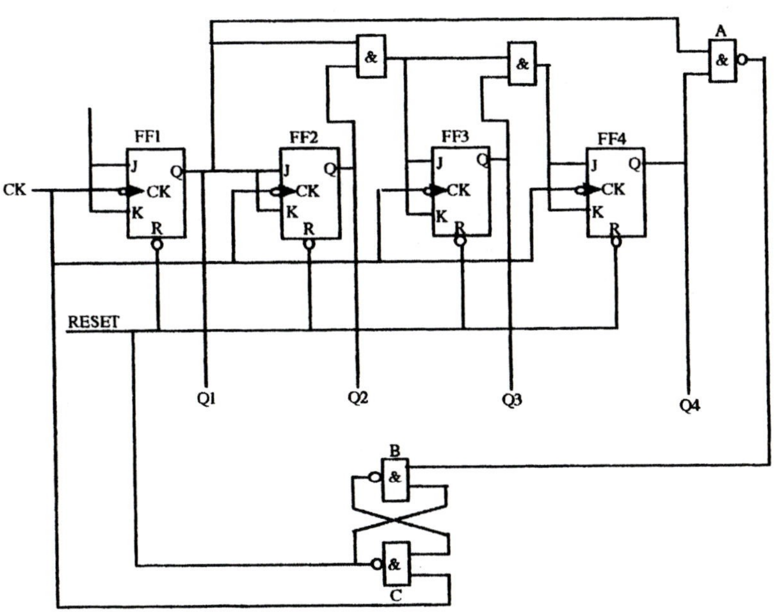

The operation of the decade counter of Figure 14.27 is exactly the same as the synchronous counter up to the point where the Q outputs go to the state 1001. The modification to the divide-by-sixteen counter starts with the addition of NAND gate A. This gate is used to detect when the counter reaches nine, 1001 in binary. At this point both its inputs, one from FF1 and the other from FF4, are logic level 1 so by applying NAND logic the output must go to 0.

This provides one input to the R–S latch gate B. The other latch input to gate C is also at 0 since there is no clock pulse present, so the R–S flip-flop goes to the forced condition with both outputs at 1. When the next clock pulse is applied the input to gate C rises to 1 and its output falls to 0 which activates the reset line and causes the output of each flip-flop to go to 0, the reset condition. FF1 does not respond to this clock pulse because the direct reset input has priority over all other inputs when activated. Gate B input now returns to 1 since gate A has a 0 on each of its inputs, while gate C input falls back to 0 because the clock pulse has ended. The R–S latch now adopts the condition with gate B output at 0 and gate C at 1. The reset line is high and no longer active. The count can then proceed until the next occasion when a 1001 code is detected.

Shift registers

The fact that a flip-flop can store information in the form of a binary 1 or 0 has by now been established. Any circuit which can be used to store information in binary digit form, *bits,* is called a register. Circuits which can not only store bits but can manipulate the data by moving it around are called *shift registers.* Most shift registers use either D type or J–K flip-flops. A simple three-stage shift register is shown in Figure 14.28.

Figure 14.28
Shift register

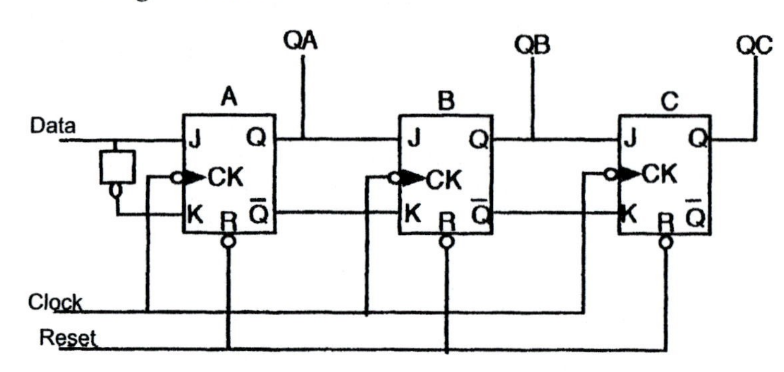

Again the best way to start to describe the operation of the circuit in Figure 14.28 is to reset each flip-flop by applying a 0 to the reset line. The data input to flip-flop A is fed directly to the J input but via an inverter to the K input. This ensures that the flip-flop output will take up the state of the J input on the trailing edge of the clock pulse. If J=1 when the first clock pulse is applied the output at QA will go to 1 on the trailing edge. QB and QC will not change because their J inputs were at 0. If the J input to flip-flop A is now returned to 0 and another clock pulse applied QA output will go to 0 while QB output will go to 1 since its J input was at 1 when the clock pulse was applied. QC remains unchanged. If the J input to flip-flop A is again taken to 1 and a third clock pulse applied its output will go to 1, QB output falls to 0 while QC goes to 1 since its input was at 1 when the clock pulse was applied. The three bits of data 101 applied at the input have now been clocked into the register by shifting action. What has been created is the electronic equivalent of a conveyor belt. Examining the action of this circuit reveals that after three clock pulses the original data 101 can be examined at the same time by looking at outputs QA, QB and QC, this is described as parallel form. Not only is the register storing the information, it is converting it from a serial form into a parallel form. The shift register can also act as a *serial to parallel converter* or *SIPO, serial in parallel out,* device. Furthermore if clock pulses are continually applied the data will continue to move from left to right and appear at QC output bit by bit. The register is also acting as a *serial in serial out* or *SISO* device.

Parallel to serial conversion

Just as it is possible to construct counters in either asynchronous or synchronous form it is also possible to do the same with data converters. In either type of converter the first step necessary is to stop the clock pulses while the data is loaded and then to prevent any further data being loaded until the clock has been able to push out the data in serial form. The circuit of Figure 14.29 is an asynchronous converter. It makes use of the direct set and reset inputs to load the data and then the clock is used to move the data through using the J and K inputs. When the last bit of data has passed out the clock is stopped again and new data loaded allowing the process to repeat. The action required in the circuit of Figure 14.29 is to stop the clock by taking the PLE line, parallel load enable, to 1. This holds the output of AND gate A at 0 because of the action of the NOT gate V at its input, so no clock pulses are present on the clock line. At the same time this PLE line also activates NAND gates B to L. If any of the inputs are at 1 the flip-flop adopts the SET condition. Examining flip-flop 1 assume that In 1 is at 1. With PLE at 1 gate B has a 1 present at both of its inputs which produces a 0 at its output while gate C has this 0 at one of its inputs and must therefore produce a 1 at its output. The gates B and C are connected to active low inputs so C has no effect but B has gone low thus setting the flip-flop and a 1 output at OUT 1. The input In 1 has been loaded into flip-flop 1. If In 1 is a 0 then gate B would produce a 1 output and gate C would produce a 0 output resetting the flip-flop causing a 0 output at OUT 1. When PLE returns to 0 one input to AND gate A is held at 1 so its output is produced by the other input which is the clock. Clock pulses now move the data stage by stage, passing it out in serial form at OUT 4.

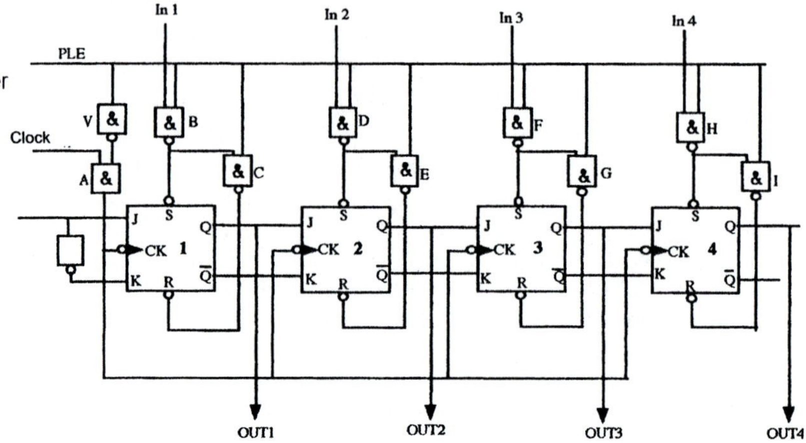

Figure 14.29
Parallel to serial converter

It is also possible with this register to load the data in four bits at a time and store it, then use the outputs OUT 1 to 4 to provide it in parallel form. To summarise, it is possible to use shift registers as storage devices, called sequential access memories (SAM), serial in serial out (SISO), serial in parallel out (SIPO), parallel in serial out (PISO) or parallel in parallel out (PIPO) devices.

Clock circuits

Already established with shift registers and counters is the need for a clock or pulse generator. Any oscillator which produces a series of square pulses at regular and constant intervals can be used as a clock. The circuit of Figure 14.30 shows a simple clock oscillator using a pair of NAND gates.

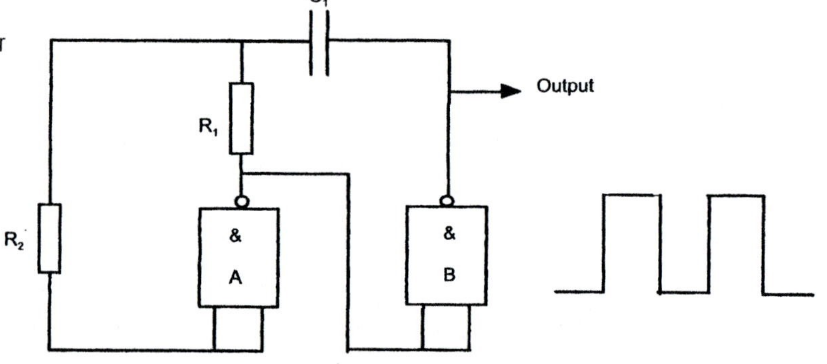

Figure 14.30
Clock oscillator using NAND gates

The action of the circuit is quite simple. At switch on assume that the output from gate A is logic 1, this means that gate B has both its inputs at the same logic 1 level while the input to gate A on both its inputs must be logic 0. The output of gate B is 0 and under this condition the left-hand plate of capacitor C_1 begins to charge through R_1. The voltage on its left-hand plate rises. Eventually it will reach a level positive enough to act as a logic 1 which is applied to gate A inputs through R_2. Gate A must now change its state and its output level falls to logic 0 changing the input to gate B and as a result its output which rises to logic 1. Under this condition capacitor C_1 now discharges through R_1 with the voltage on its left-hand plate returning to 0. The input to gate A is once again logic 0 so its output goes to logic 1 and starts to recharge the capacitor. The action is repeated at a rate determined by the rate of charge and discharge of R_1 and C_1. Since both of these components have a working tolerance it is quite likely that their charge/discharge rate will vary and the output frequency will change. Clearly if the purpose of a clock is to generate timing signals accurately this circuit is not good enough. Where accurate timing is required it will be necessary to use an oscillator controlled by a crystal. A typical circuit of a crystal controlled clock oscillator is shown in Figure 14.31.

Figure 14.31 Crystal oscillator using 74LS04 TTL gates

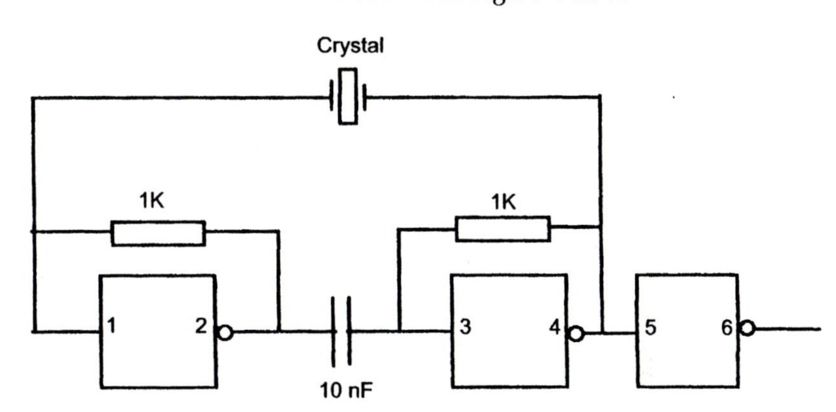

It would not be possible to produce a crystal oscillator for low frequency operation. One problem, that of stability, has been changed for another. However, if a high frequency is chosen for the oscillator operation using the techniques described earlier in the chapter it is possible to reduce the output frequency by using a counter/divider circuit to obtain the desired frequency.

The 555 timer

This is a small eight pin DIL integrated circuit. When powered from a +5 V and ground supply it can be used with either TTL or CMOS logic gates as a clock oscillator producing a continuous train of pulses. Figure 14.32 shows the pin connections of a 555 timer IC and a typical circuit connected to perform as an astable multivibrator.

Figure 14.32
555 astable multivibrator

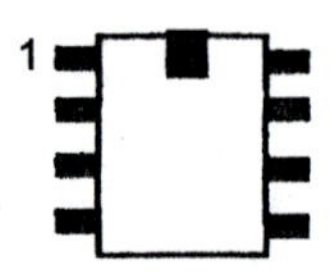

Pin 1 ground
Pin 2 trigger
Pin 3 output
Pin 4 reset
Pin 5 control voltage
Pin 6 threshold
Pin 7 discharge
Pin 8 supply +V

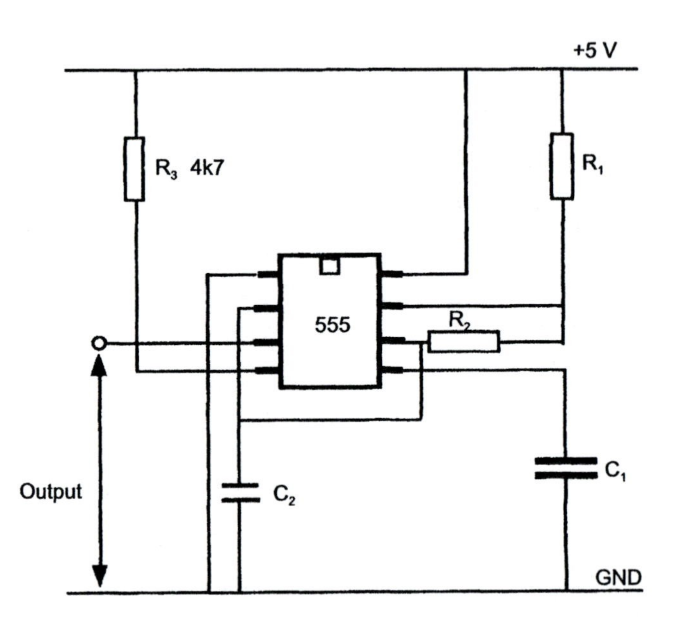

The 555 circuit operates as follows. When the supply is connected the output at pin 3 is at +5 V, the high state. The capacitor C_1 connected to pin 2 and acting as a timing circuit begins to charge through the two series resistors R_1 and R_2 until the voltage is equal to 66% of the supply voltage. Inside the integrated circuit a voltage comparator senses when the voltage has reached this level and when it has it causes the output to change state so that it now goes low, 0 V. This action discharges the capacitor C_1 through R_2 to ground. A second voltage comparator is used to detect when the capacitor C_1 has discharged to 33% of the supply voltage and when it has it switches the output back to the high state, turning off the discharge transistor and restarting the cycle. As can be seen later on in this chapter the two voltage comparators are used to drive the R and S inputs of an R–S flip-flop. The frequency of the waveform generated can be determined by using the following formula:

$$f = \frac{1.44}{(R_1 + 2R_2)C_1}$$

The values chosen for R_1 and R_2 also determine the shape of the output waveform. For a squarewave output with a mark to space ratio of 1:1 R_1 needs to be much smaller than R_2 with R_1 equal to 1 kΩ and R_2 equal to 47 kΩ. Alternatively, if R_2 is replaced by a variable resistor the output frequency can be made variable but R_1 must remain fixed at about 1 kΩ.

The 555 timer as a monostable

This device can also be used to act as a monostable multivibrator producing an output pulse only when a triggering pulse is applied at the trigger input pin 2. This triggering pulse must be a negative going pulse changing its level from the high state to the low state. The time duration of the trigger pulse must be a shorter time period than the output pulse.

Figure 14.33
555 monostable multivibrator

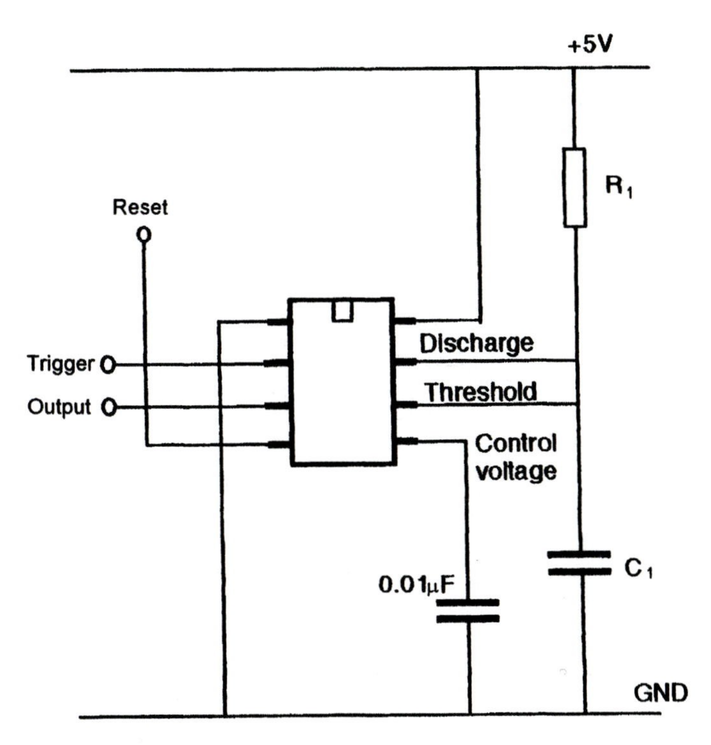

The pulse duration for this circuit can be determined by the following calculation:

Pulse duration $= 1.1\ R_1 \times C_1$

If the value of $R_1 = 1$ kΩ and the value of $C_1 = 0.01$ μF the duration of the pulse will be 11 μs. It is important to keep the connection to the trigger input as short as possible to avoid false triggering caused by induced noise pulses. For pulse durations in the high output condition for times less than 2 or 3 μs an alternative device will have to be used. Figure 14.34 shows in simple form the internal block diagram of the 555 IC. To substitute the IC by using discrete components would require a large circuit board with a considerable number of components.

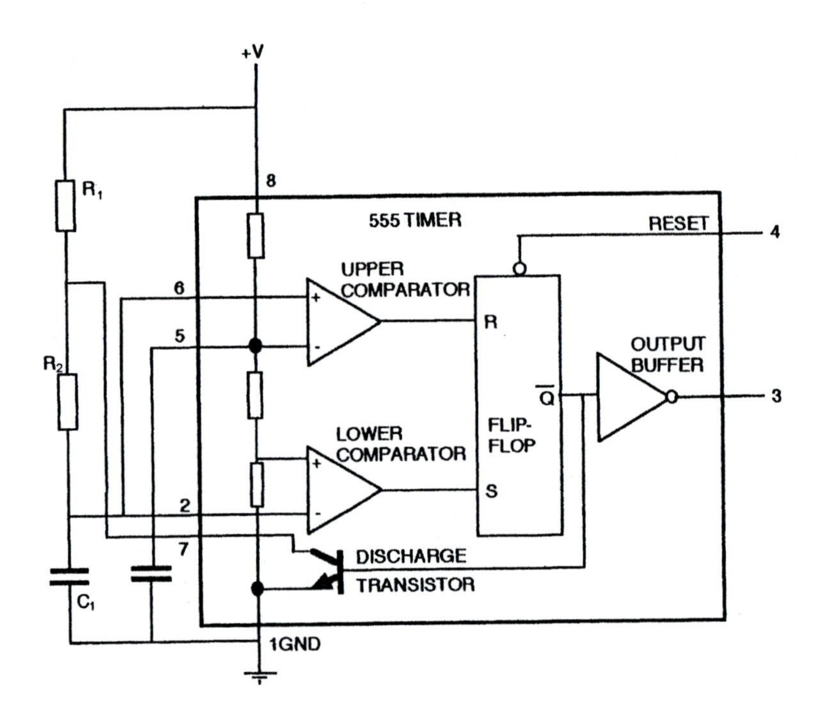

Figure 14.34 Internal block diagram of a 555 timer

Interchanging logic gates

When examining logic gates earlier in this chapter a typical logic gate was shown (Figure 14.12), a 7400 quadruple NAND device. This gate contain four identical but separate logic gates inside one integrated circuit. It would be inconvenient for a circuit designer to have to include in his circuit more integrated circuits than was necessary and to leave some gates in certain integrated circuits unused. To avoid this and make circuits as efficient as possible experimentation shows that it is possible to obtain an OR function using NAND gates or indeed any other logic function required. When examining the circuit of the clock oscillator in Figure 14.30 a NAND gate was used with both its inputs connected together. In this form the NAND gate produces a NOT function and examining the truth table and diagram in Figure 14.35 reveals why.

Figure 14.35
The NOT function using a NAND gate

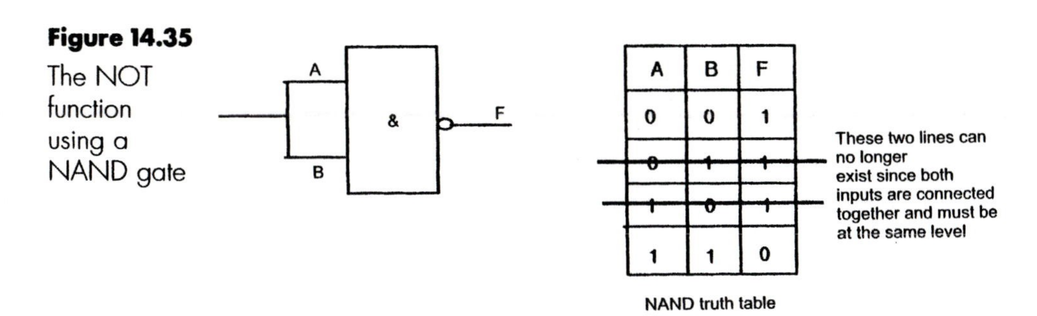

Just as Figure 14.35 shows that NAND can become NOT the following series of diagrams in Figure 14.36 reveals that interchanging one type of logic gate device with a different type of logic gate suitably connected can produce the same logic function.

Figure 14.36 shows that by inverting the inputs to a NAND gate (by the use of two other NAND gates connected as NOT gates) it is possible to produce the same output as an OR gate. This could easily be converted to a NOR gate by inverting the output through another NAND gate connected as a NOT gate. Similarly it is possible to obtain the AND or NAND function by using a series of suitably connected NOR gates. This is demonstrated in Figure 14.37.

Figure 14.36
NAND gates used to produce the OR/NOR function

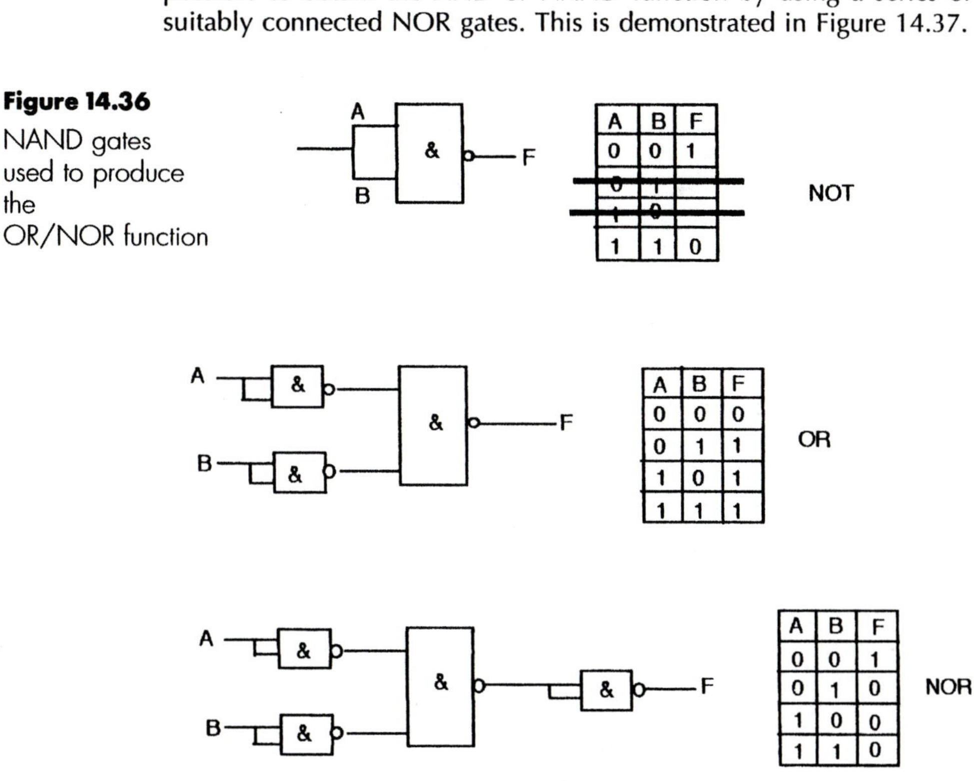

Figure 14.37
NOR gates connected to produce the AND/NAND function

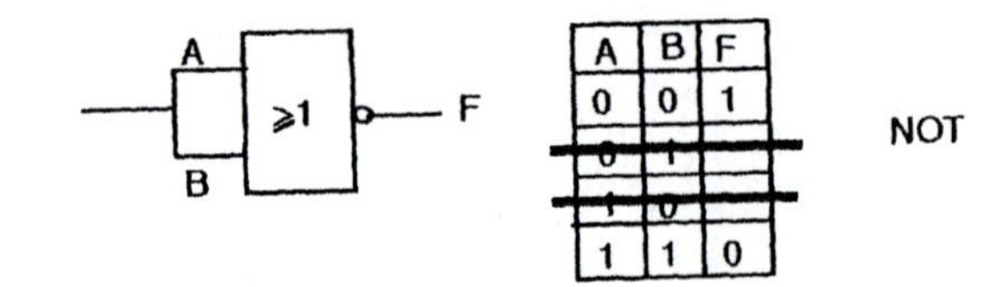

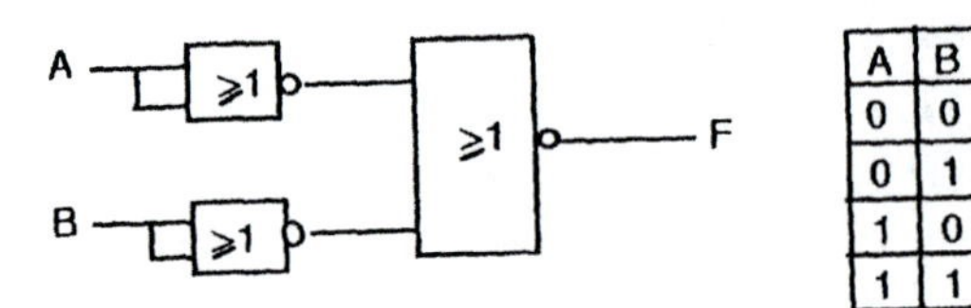

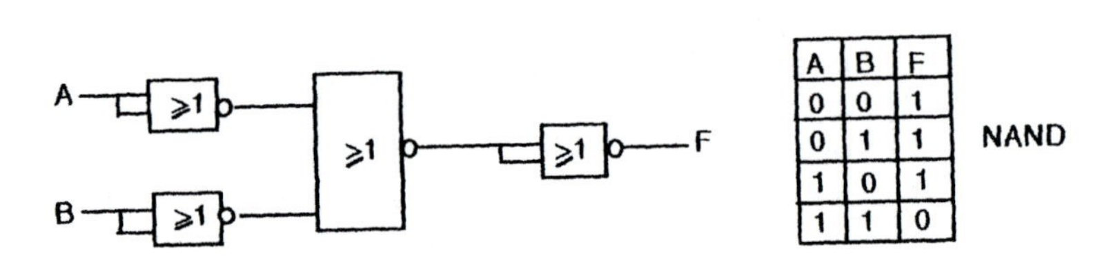

15

Test equipment

The complexity of modern electronic circuits demands sophisticated and accurate test equipment, and much valuable time can be saved in both fault location and alignment when the correct items of test equipment are available. On the other hand, equipment alone will not locate the fault; only when it is in the hands of a competent engineer will the value of the test equipment be appreciated, so it is essential that all engineers are conversant with the operation, applications, and limitations of the test equipment at their disposal.

The multimeter

The multimeter is perhaps the most versatile piece of test equipment at the service engineer's disposal. It can be used for measurement of both dc and ac voltage and current in circuits, as well as testing of circuit resistance and continuity. It can also verify the condition of a number of types of component.

Two types of multimeter are available: *analogue* and *digital*.

The analogue meter dates back many years. It functions by passing a sample of current from the circuit under test through a fine coil of wire, hence the name *moving coil meter*. The meter movement comprises a permanent magnet and a pivoted coil with an attached needle. When a sample of current from the circuit under test is passed through the coil, a magnetic field is established. This interacts with the permanent field causing the pointer to deflect. The degree of deflection is proportional to the force acting upon it, which in turn is proportional to the sample current.

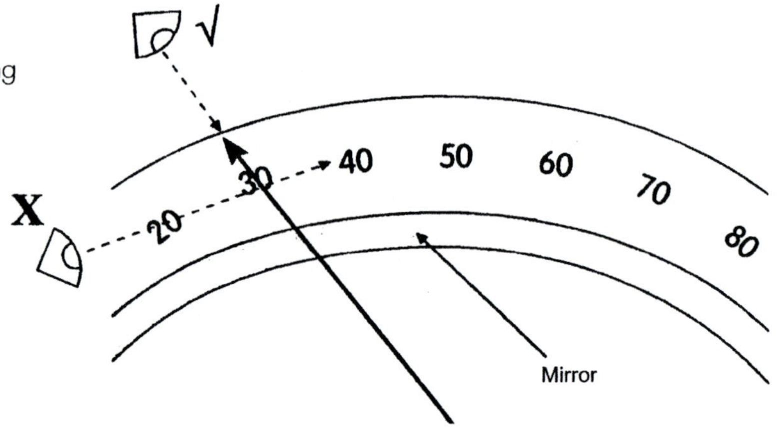

Figure 15.1 Incorrect reading caused by parallax error when pointer needle is not sighted using the mirror

A good quality analogue meter provides very accurate readings, although care must be taken when using the meter to read it looking vertically down, and not from the side, otherwise a parallax error will occur (see Figure 15.1). To assist accurate reading a mirror is usually placed on the scale behind the pointer. The reading is taken when the needle has been sighted directly above its reflection in the mirror.

Analogue meters are particularly sensitive to small but rapid fluctuations in current or voltage, which makes them an ideal choice when performing adjustments to electronic equipment. The main disadvantage with the analogue meter is the very delicate precision movement which cannot withstand any severe shock. This does not make it the ideal choice for field service. Another problem associated with the movement is the fact that the coil has a maximum current rating of 50 μA. If there is no protection circuit the meter is very easily damaged if accidentally overloaded.

The digital meter has largely overtaken the analogue type in popularity, mainly because it is more robust (no moving parts). Other advantages are: there is no problem with parallax error, they are comparatively less expensive than analogue types of equivalent performance, and many versions are available with additional features such as built-in capacitance meter, frequency counter, thermometer, diode and transistor tester, continuity buzzer, and other useful devices.

The only drawback with many digital meters is that they sample the voltage or current they are reading at intervals of up to 0.5 seconds. This makes it difficult when carrying out adjustments in circuits where a continuous readout is desirable. However, some digital meters offer a near continuous analogue readout in the form of a bar display to help overcome this problem, a feature that is worth spending the extra on when purchasing a meter.

All multimeters should be treated as a delicate instrument; for that is what they are. But apart from careful handling, careful use, i.e. correct operation, is also paramount. A meter can suffer serious damage if it is either set to the incorrect range, or if it has been connected the wrong way around. Therefore, the first rule when using a meter is to *ensure that it has been set to the appropriate range before connecting it to the circuit to be tested.* The second rule is to ensure that, where appropriate, *the polarity of the meter connections to the circuit is correct.*

The three most common tests a service engineer makes with a multimeter are voltage, resistance, and current.

An example of a voltage test is given in Figure 15.2 where a digital meter is shown connected across resistor R_2. In this case the meter will indicate the p.d. across the resistor. Note that the meter has been set to a *dc* voltage range because that is the nature of the voltage expected across R_2. Also, the range is higher than the supply, thus ensuring that the meter will not be overloaded. Once connected, if the meter reading is found to be less than 2 V it can be switched to the 2 V range to obtain a more accurate reading.

Also note the polarity of the leads in relation to the battery polarity. The positive lead is connected to the positive end of R_2. In the case of a digital meter, incorrect polarity in this case would most probably not result in damage to the meter, it would simply indicate a negative voltage on the readout. However, in the case of most analogue meters incorrect polarity would cause the needle to attempt to move backwards, which could well result in permanent damage to the meter movement.

To measure current the meter must be connected in series with the circuit in order that the current flowing around the circuit passes through the meter. To do this the circuit must be broken at some point, which often means desoldering a lead or one end of a component. In the example shown in Figure 15.3 the circuit has been broken by disconnecting the positive battery lead.

The ac voltage and current ranges are intended for measurement of low frequency (50 Hz mains) sources. Attempts to measure h.f. audio or radio frequency signals on these ranges will result in an incorrect reading.

Extreme care must be taken when using the meter to measure ac mains potentials (or even greater). An accidental short circuit, or application of the meter to the mains terminals whilst set to the incorrect range, can result in the leads melting or the meter exploding which can cause serious burns to the engineer. To offer a degree of protection against such mishaps the Health and Safety Executive insist that probes which are to be used for testing high voltages must have finger barriers to protect against inadvertent contact with the terminal, and must have no more than 4 mm of uninsulated metal at the probe end. Furthermore, the probe must have an in-built fuse of 500 mA maximum rating. The leads for such probes must be adequately insulated and sheathed, and should be clearly coloured red and black.

Figure 15.2
Voltage measurement across R_2

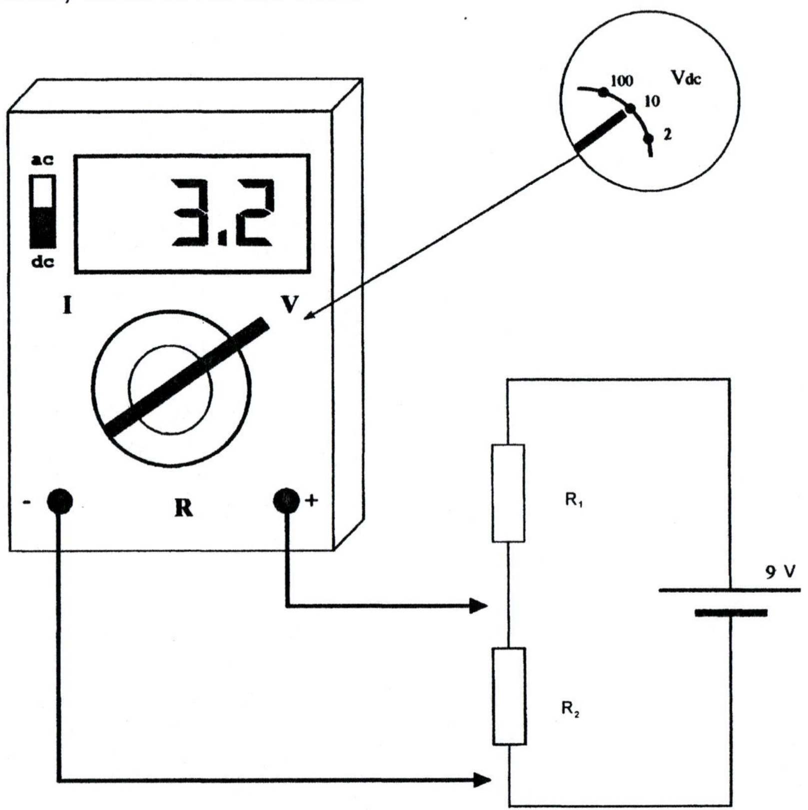

Resistance measurement differs from voltage and current in that the meter is not extracting a sample of current from the circuit, but rather, by utilising its internal battery, it passes a sample of current into the circuit or component to be tested, and calculates the resistance by measuring the amount of current that flows from its known battery potential. The important point to note is that there must not be any other voltage present across the component to be tested as this could cause serious damage to the meter, and will in the very least give an inaccurate reading.

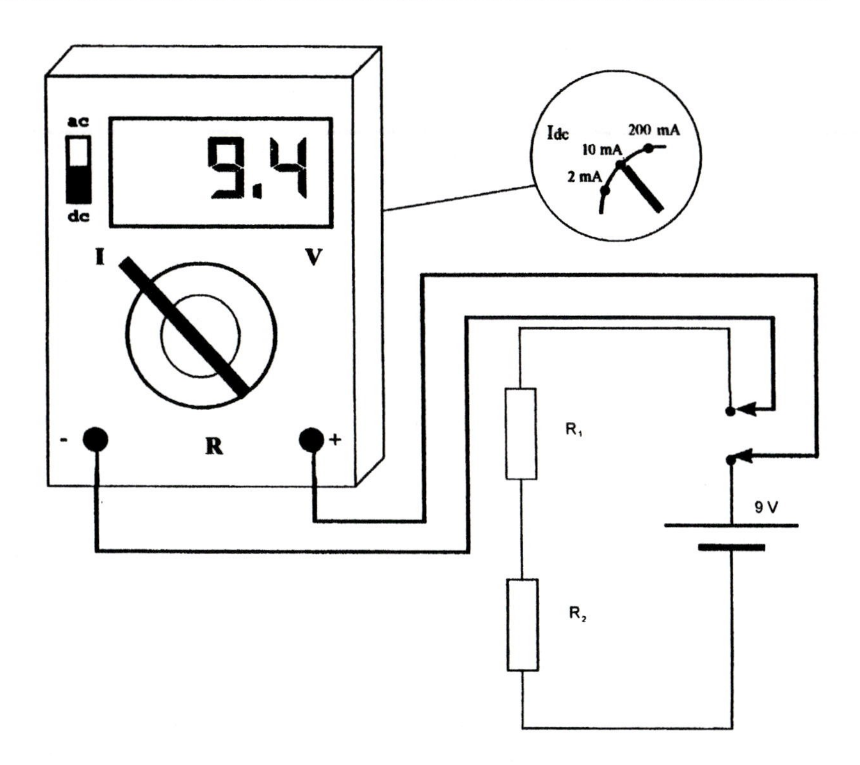

Figure 15.3 Measuring total circuit current

When measuring the resistance of components it is good practice to disconnect them from the circuit before carrying out the test. This not only ensures that there is no other voltage present, but also gives a more reliable reading as there will not be any other components connected in parallel with the one you are trying to test.

Methods of testing some common components are shown in Figure 15.4.

In Figure 15.4(a) a carbon resistor is being tested. In this case the polarity of the meter leads is not important; however, the resistance range selected must be higher than the value of the resistor. When set to a high range, make sure that you do not place your fingers across the probes, otherwise the meter will give an incorrect reading due to the parallel resistance of your body.

In Figure 15.4(b) an inductor is connected to the meter which is switched to a low resistance range. The low range is selected because the meter will only read the dc resistance of the wire, which will be no more than a few hundred ohms at the most, and in the majority of cases will be just a few ohms. This same test applies to other low resistance devices such as fuses, which should show a value of almost 0 Ω, and filament lamps which normally have a resistance of a few ohms.

Figure 15.4

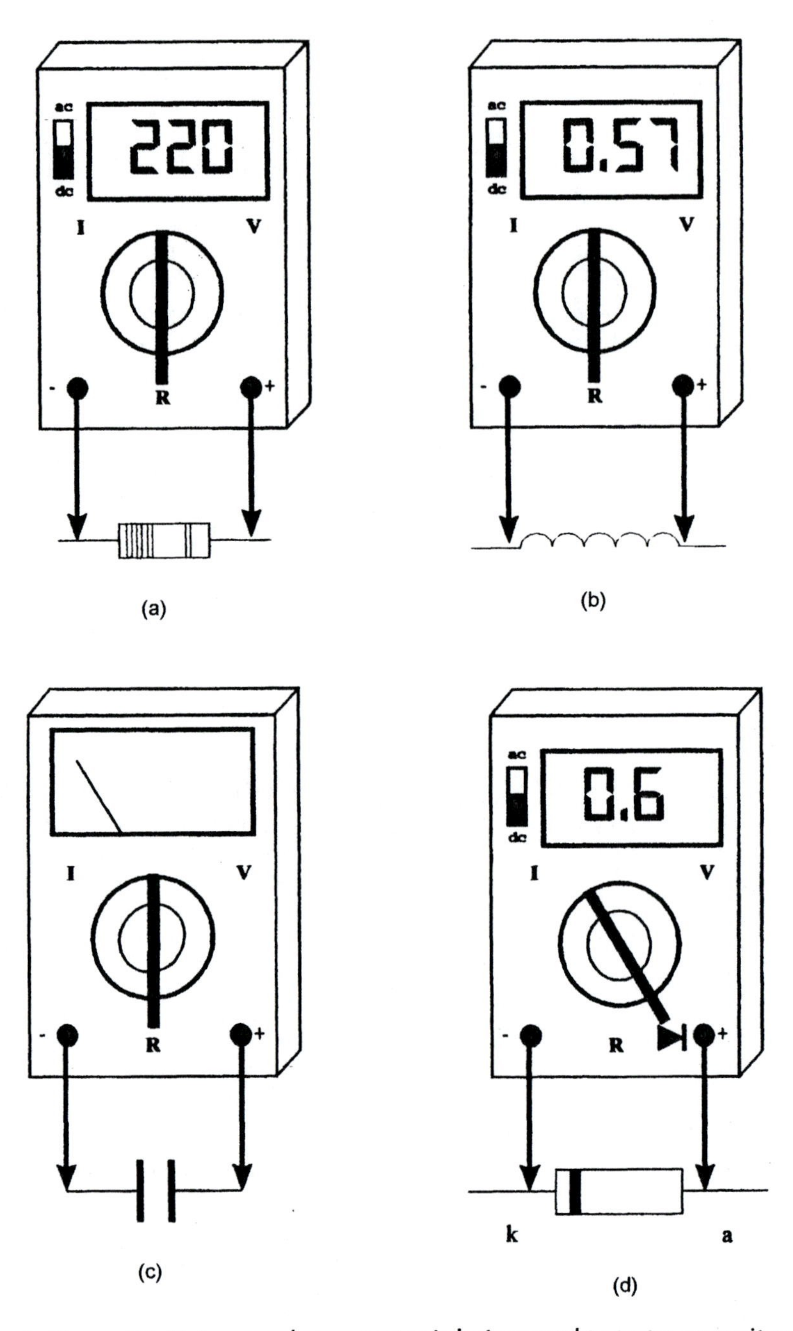

In Figure 15.4(c) an analogue meter is being used to test a capacitor. There is no way that the value of a capacitor can be ascertained using the ohms range of a meter; however, if the capacitor is suspected of having gone short circuit or leaky (that is, resistive) the meter will confirm this. With larger values of capacitor (100 nF and above), if the device is good the meter needle will kick at the moment the leads are applied as the capacitor charges. If the leads are reversed there will be another kick as the capacitor charges in the opposite direction. This kick only verifies that the device has not gone open circuit, it does not really indicate if the value is correct. This test does not work as well on a digital meter due to its sampling the charge current. You simply have a series of numbers flashing for a moment which, although they indicate that charging has taken place, you find difficult to read.

Diode testing (Figure 15.4(d)) on a digital meter cannot be performed on the resistance range because the meter does not output a high enough test voltage to bias the diode into conduction. For this reason digital meters incorporate a *diode test* facility. When the meter is connected positive to anode and negative to cathode, the diode conducts and the meter gives a reading of the junction volt drop which is around 0.6 V for silicon, 0.2 V for germanium, and between 1.2 V and 2.3 V for an LED (depending on the colour of the light output). When the leads are reversed the diode will not conduct, and the meter will give an open circuit indication.

When using an analogue meter to test a diode, a forward resistance reading in the order of 800–1.2 kΩ is typical for silicon and gallium/arsenide (LEDs), and 100–200 Ω for germanium.

When a meter is connected to test voltage, it can have an effect on the circuit under test because a small current flows from the circuit into the meter. Clearly, from the point of view of testing, the less current taken from the circuit the better, so by design the meter should have a high internal resistance. On the other hand, the meter must be able to draw enough current to enable it to give an accurate measurement. The ideal meter is one that has a very high resistance but is able to give highly accurate readings from the resulting low sample current.

In practice only the more expensive meters are able to come close to the ideal. Many budget meters require a relatively large sample current to take a reading, which, as we shall see in a moment, can result in inaccuracies.

The *sensitivity* of a meter is measured in *ohms per volt*. This figure refers to the ohmic value of every volt to be measured. For example, let us suppose that a meter requires 50 μA to give the maximum reading on any particular range. Thus, using Ohm's law, the resistance of the meter to measure 1 V would be:

$$R = V \div I = \frac{1}{50 \times 10^{-6}} = 20\ \text{k}\Omega$$

Therefore we say that the meter has a sensitivity of 20 kΩ/volt. Thus, on the 10 V range the meter has a resistance of 10 × 20 000 = 200 kΩ, but when set to the 100 V range this rises to 2 MΩ.

Let us compare this meter with another that requires 1 mA to give the maximum reading on a range. This meter would have a sensitivity of:

$$R = \frac{1}{1 \times 10^{-3}} = 1\ \text{k}\Omega/\text{volt}$$

Referring to Figure 15.5, we shall now examine the effect that each of these meters has on the circuit, and the resulting readings obtained.

Figure 15.5

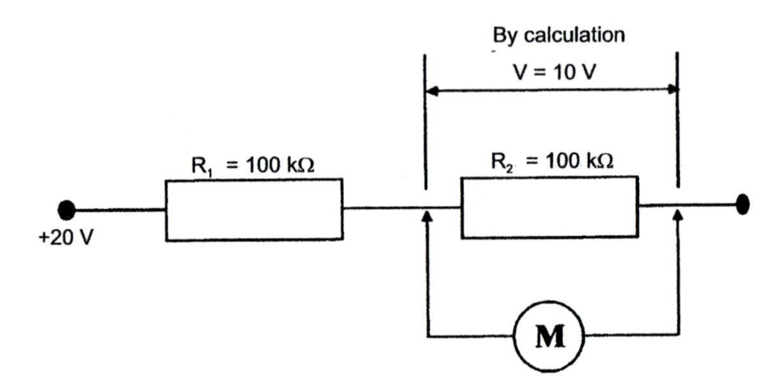

We know that the value of VR_2 must be 10 V. If the meter with a sensitivity of 20 kΩ/V is connected across R_2 whilst set to the 10 V range, then the effective value of R_2 changes to 100 k//200 k = 67 kΩ. Hence the meter would give a reading of 8 V.

In the case of the 1 k/V meter (when used on the 10 V range) R_2 now becomes 100 k//10 k = 9.1 kΩ. Hence this meter would give a reading of 1.7 V!

The problem of the meter loading the circuit is most acute when measuring voltages in the proximity of high circuit resistance values, and the service engineer must be aware of this, otherwise they could be misled by the incorrect readings obtained.

The oscilloscope

The cathode ray oscilloscope (CRO) is an essential piece of test equipment in any modern service department. Although used primarily to display ac signal waveforms, allowing them to be measured and analysed, the CRO can be used to measure dc voltages as well. The main advantages over the multimeter are its high input impedance, that is, it does not exhibit the loading problems we have just seen with meters, and the fact that it can measure ac signals with frequencies far in excess of the few kilohertz range of the multimeter.

A typical front layout of a double beam scope is shown in Figure 15.6, and a simple block diagram is given in Figure 15.7.

Figure 15.6
Typical basic dual beam oscilloscope

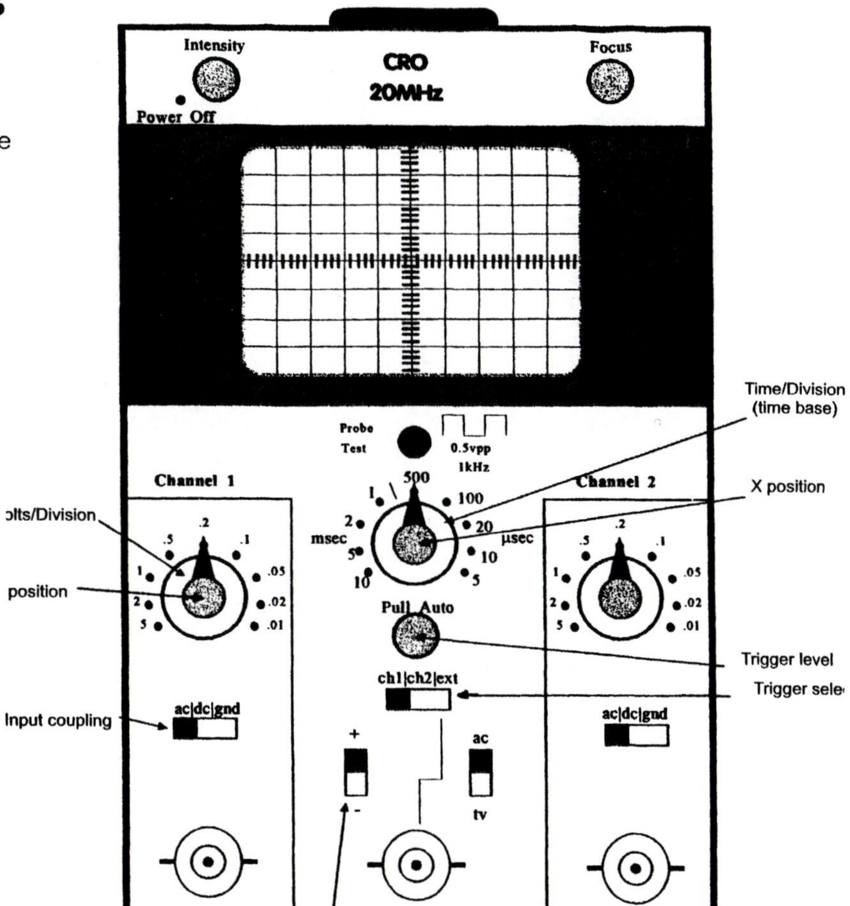

The CRO produces a graph of voltage against time, with time on the *x* or horizontal axis. The screen is covered with a grid of squares which are usually 1 cm square, this grid is termed the graticule. There are subdivisions on the two-centre horizontal and vertical lines to permit more accurate readings to be taken if required.

Inside the scope is a *time base* that deflects the electron beam horizontally. The speed of this time base can be altered using the time base control which can produce anything from a single spot moving slowly across the screen, to a horizontal line; or two lines if both Y channels are turned on.

The X and Y position controls enable the beam(s) to be moved horizontally and vertically around the screen to provide suitable separation between them, as well as to permit a waveform to be moved to a position on the graticule where an accurate measurement can be taken.

When a signal is connected to a Y input, the beam will deflect vertically in sympathy with the signal. The larger the signal the greater the amount of deflection. Because the beam is now moving both horizontally and vertically, it will plot a trace on the screen that represents the Y input waveshape. Of course, the correct waveshape will only be obtained when both time base and volts/division controls are set to values that relate to the periodic time and peak to peak voltage of the input signal.

A clear graph will only be obtained if the beam begins to plot the input at the same point on every horizontal scan of the CRO screen. To ensure that this happens a *trigger* control is provided which synchronises the time base to the incoming signal. Where the input signal is a relatively non-complex waveshape, i.e. a sine wave or squarewave, the trigger control can be set to 'AUTO' and the 'scope will find its own suitable trigger point. However, where more complex signals such as television luma, chroma or video waveforms are being displayed, the trigger may have to be set manually.

Where two input signals are being displayed, one on each channel, the signal to which you wish the 'scope to trigger is selected using the 'TRIGGER SELECT' switch. Alternatively, the 'scope can be triggered to a third input signal that is connected to the 'EXT TRIG' input, although this signal is not displayed on the screen.

The trigger +/– switch allows the engineer to select whether the 'scope triggers to the rising (+) edge of a waveform, or the falling (–) edge.

The input coupling switches on the Y inputs have three positions. The 'ac' position switches a capacitor in series with the input socket, thus blocking any dc components that would otherwise enter from the circuit under test. The 'dc' position bypasses this capacitor enabling the 'scope to measure dc voltages. Position 'gnd' means ground. This connects the input to earth and prevents any input signal, either wanted or otherwise, from entering the 'scope. This is used when setting the beams to an exact position before performing measurements.

The probe test facility is used to ensure that the probes, leads, and 'scope are all functioning correctly, and are properly calibrated before carrying out any measurements. This test point gives a known accurate signal, i.e. 1 kHz squarewave at 0.5 Vpp.

A simple block diagram of a single beam CRO is given in Figure 15.7. This can be related to the controls and inputs in Figure 15.6.

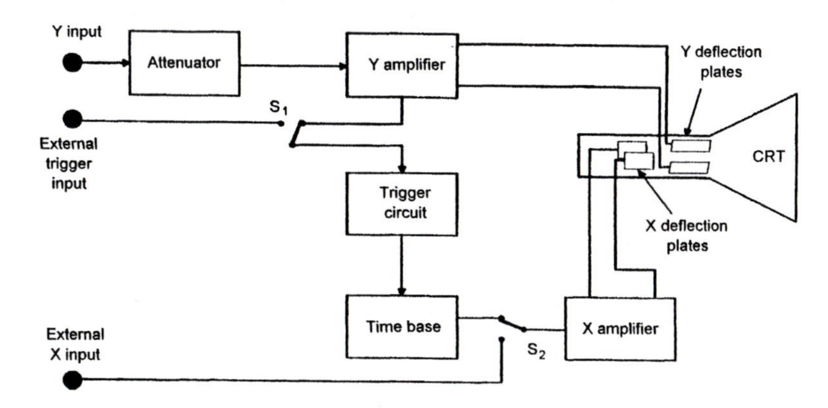

Figure 15.7 Block diagram of a cathode ray oscilloscope

The *attenuator* is a switched variable resistance which is set by moving the 'Volts/Division' control on the 'scope. This ensures that the signal at the output from the attenuator will be at the correct amplitude to fit onto the 'scope screen.

The cathode ray tube in a 'scope uses electrostatic deflection to enable the high scanning speeds required. The function of the *Y amplifier* is to boost the amplitude of the Y input signal to a sufficiently high voltage level to drive the Y deflection plates.

The triggering function described earlier is performed in the *trigger circuit* block. This circuit is set, using the trigger level control, so that the time base begins each sweep of the screen when the Y input is at the same point. Switch S_1 is the trigger select switch which enables the time base to be triggered from a signal other than that at the Y input. This is used when it is necessary to analyse a portion of a complex wave, or where the complexity is such that neither automatic or manual triggering is able to lock the trace steadily on the screen.

The *time base* is an oscillator which generates the horizontal scan and flyback drive signal. Its frequency can be altered using the 'Time/Division' control. The output from the time base is boosted in the *X amplifier* before being passed to the X deflection plates. Switch S_2 (not shown in Figure 15.6) enables the X plates to be driven from an externally generated deflection signal. Although not frequently used, this feature does have some practical applications, e.g. when the 'scope is being used to calibrate other test equipment.

Measurement of the waveshape is performed by counting the number of horizontal and vertical divisions (and subdivisions) occupied by *one cycle,* and multiplying by the time and voltage control settings. An example is shown in Figure 15.8. In this instance one cycle can be conveniently taken as being between points A and E, or points B and F.

Figure 15.8

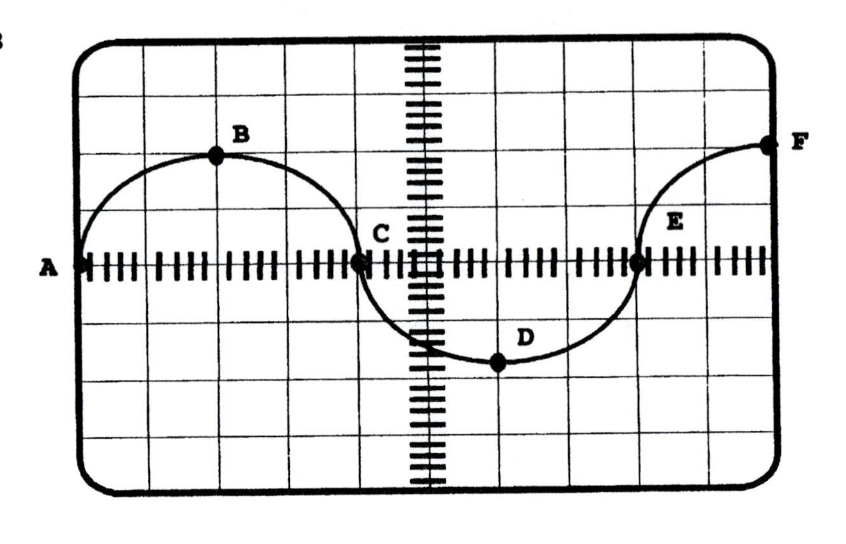

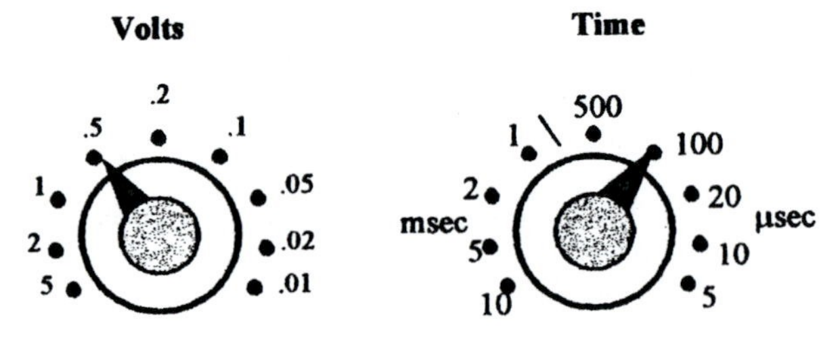

The peak to peak voltage is measured by counting the number of vertical squares between points D and B, which in this case is 4, and then multiplying this figure with the volts/division control setting, in this case 0.5 V. So the peak to peak voltage is $4 \times 0.5 = 2$ Vpp.

The frequency of the waveshape is a little more involved because you cannot measure frequency straight off a CRO screen, you can only find the periodic time. The frequency is then found using the formula $f = 1/t$. For the example in Figure 15.8, the periodic time is found by multiplying the number of horizontal squares between points A and E (which is 8) by the time base control setting (100 μs). Thus periodic time $= 8 \times 100\ \mu s = 800\ \mu s$. The frequency is now found by taking the reciprocal of the periodic time:

$$f = \frac{1}{800 \times 10^{-6}} = 1250 \text{ Hz or } 1.25 \text{ kHz}$$

Despite the high impedance of the CRO inputs, there are occasions where loading of the circuit can still pose problems. There are also occasions where the peak to peak value of a signal is so large that it will not fit onto the display even when the volts/division control is set to its highest position. In such cases a probe with a known in-built impedance can be used. The impedance in the probe is effectively in series with the signal and both increases the input impedance of the CRO and reduces the peak to peak amplitude of the incoming signal. However, this now means that the reading on the CRO will not be correct because, as shown in Figure 15.9, if the actual amplitude is 10 Vpp and a probe with a voltage division factor of 10 is used, then the signal entering and subsequently displayed on the CRO will only be 1 Vpp. But if the engineer knows that the probe has a division factor of 10, the correct reading can still be taken by multiplying the CRO readout of 1 Vpp by 10 to attain the true value of 10 Vpp. For this reason the probe is referred to as a *times ten* (×*10*) probe because the engineer must multiply the CRO readout by 10. Many probes are switchable between ×1, ×5, ×10.

Figure 15.9

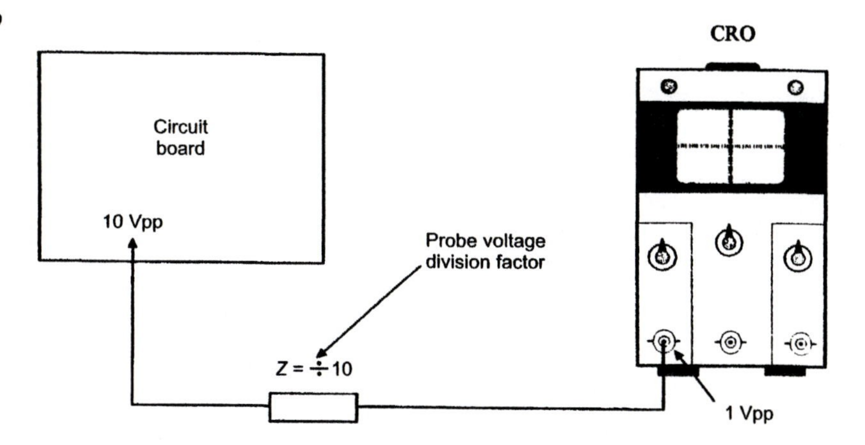

Some oscilloscopes have a facility to store in digital memory a single sample of the waveshape. This can then be displayed continuously, rather like a still frame on a video recorder, allowing careful examination and analysis. These 'scopes are known as *storage 'scopes*. Some storage 'scopes have a built-in printer that makes available a hard copy of the display.

Whilst learning to use the 'scope, individuals can spend a lot of time simply trying to set it up and obtain a stable trace. There are a wide range and variety of oscilloscopes, so it is not possible to provide a foolproof method to initialising a 'scope; however, the instructions below should be applicable to the majority of basic double beam oscilloscopes and are intended as a guide.

1. Switch on
2. Adjust INTENSITY (brightness) control to mid position
3. Set TRIGGER LEVEL control to AUTO position
4. Set channel 1 Y POSITION control to the mid position
5. Set channel 2 Y POSITION control to the mid position
6. Set TRIGGER SELECT switch to CH1 position
7. Set trigger switches to AC and + positions
8. Set INPUT COUPLING switches on channels 1 and 2 to AC position
9. Readjust ch1 and ch2 Y POSITION controls so that each horizontal line is in a different position on the screen
10. Adjust FOCUS and INTENSITY controls for best definition

The power supply

This is invaluable when servicing low voltage dc equipment as it takes away the need to insert batteries, something that can prove difficult once a piece of equipment has been stripped down for servicing. A bench power supply can also be useful in fault diagnosis when a power supply rail in a piece of equipment has failed; the bench supply can be used as a temporary substitute whilst the serviceability of the rest of the circuits is ascertained.

Although a power supply with a fixed output voltage is of some use to the service engineer, a variable output makes it far more versatile. An output between 0 V and 30 V at 1 A maximum is typical, although a higher current may be required for some applications.

Any reasonable quality power supply will have a current overload protection circuit that will cause the unit to shut down either in the event of the circuit under test drawing a current greater than the maximum rating of the power supply, or the output leads becoming temporarily short circuited.

Apart from being able to vary the output voltage, some units have a variable *current limit* facility. This allows the engineer to decide on the maximum current a circuit can draw before the power supply shuts down, thus giving a degree of protection to a faulty circuit that would have otherwise drawn too high a current resulting in further damage.

The function generator

Sometimes referred to as a signal generator, this item of test equipment is used to generate voltage waveform signals of differing amplitudes, shapes and frequencies.

Many electronic circuits are designed to process signals; however, under test conditions these signals are not always available. For example, a domestic audio power amplifier is designed to amplify audio signals from a cassette deck, CD player, etc., but it is very difficult to test such an item in a workshop when it has been returned for service on its own. Also, to perform certain tests on such an amplifier a constant sinusoidal input is required, thus a constantly changing music signal waveform is of no use. Likewise when servicing a television or VCR there are occasions where a constant display is required. Such a display is only available from a suitable pattern generator.

There is a range of generators available covering differing applications. Low frequency (l.f.) units generally offer a range of output frequencies somewhere in the order of 15 Hz–1 MHz, with the facility to select sinusoidal or square waveshape. An additional TTL output is sometimes provided which gives 5 V squarewave signals that can be used specifically for testing logic circuits employing TTL ICs.

Radio frequency signal generators are available which can provide both amplitude and frequency modulated signals over a wide range of carrier frequencies. They are used specifically for fault location on radio equipment; however, because of their high cost they are rarely seen in domestic equipment service departments, unless it happens to be an old model that has been around for many years dating back to when radio servicing was an everyday occurrence. RF generators are still used by engineers employed in the servicing of commercial and military radio transmitting and receiving equipment.

We have already mentioned the television pattern generator. This is an essential piece of test equipment in any TV/VCR service department as it can generate a range of picture displays which are useful in both fault finding and setting up/adjustment procedures. The output can be either on a modulated UHF carrier that can be received via the aerial input socket on the TV/VCR, or a 1 Vpp composite video signal containing both luminance and chrominance information, with a separate sound output. The most common displays are the colour bars, grey scale bars, crosshatch, and red or white rasters, although other displays are available on more elaborate generators.

Many function generators have a meter that indicates the voltage output. Others enable the output voltage to be selected using a combination of controls and switches. In either case it should be noted that these indicators are only accurate when the input impedance of the circuit under test is equal to the output impedance of the generator, which is usually 600 Ω. In most instances this will not be the case, and thus it is essential that the voltage output is verified by using an oscilloscope before carrying out any tests.

The frequency counter

As the name implies, this piece of test equipment is used to measure the frequency of sinusoidal or squarewave signals, and although this can be done using an oscilloscope, the frequency counter is far more accurate, especially at higher frequencies.

The counter operates by sampling the signal for a precise period of time (known as the *gate period*), calculating the frequency from the number of cycles counted during this time period. The frequency is then displayed whilst the counter takes the next sample, after which the display is updated with the new result. Clearly, the longer the sampling period the more accurate will be the calculated result, but if the sampling period is too long the engineer will have to wait for the display to update. This waiting period can be inconvenient when carrying out adjustments.

A gate period of around 100 ms is generally accepted as a suitable compromise between accuracy and speed when performing adjustments on equipment, whilst periods between 1 sec to 10 sec are often used to check the accuracy of the adjustment once it has been performed.

In addition to measuring frequency, some counters can measure the periodic time of the signal.

The insulation resistance tester

Traditionally this piece of equipment has been employed more frequently by electricians than electronics engineers as it is used to test the integrity of the insulation of cables and components such as transformers. It does this by applying a high dc voltage across the insulator under test. Any weakness in the insulator is revealed by a current flow, which is indicated as a resistance reading on the display.

The insulation is tested by applying a voltage of at least twice that of the normal operating voltage; i.e. for a 230 V domestic wiring circuit a test voltage of 500 V is applied. The tester usually offers ranges of 250 V for testing 110 V circuits or equipment, 500 V for 230 V equipment, and 1 kV for 415 V three phase equipment.

Portable appliance testing (PAT) is now a requirement of the *I.E.E. Code of Practice*. This requires that all electrical and electronic equipment should be tested both on a regular basis, and following any kind of repair or service. The test should include insulation resistance and earth continuity. Although these can be carried out using an insulation resistance tester, specially designed PAT testing equipment is available which simplifies the test procedure.

The logic probe

Logic probes are a simple and very useful hand-held item of test equipment employed in the testing of logic circuits. Inexpensive probes are available from any popular high street electronics supplier, or alternatively there are many circuit designs circulating for those who would prefer to construct their own versions. More expensive probes are available which can operate at frequencies which would normally require a high quality oscilloscope or logic analyser.

In most cases the simple probe will indicate whether the point to which it is connected is at a logic high (binary 1), or logic low (binary 0) state. However, it must be pointed out that logic probes are not universal. They are usually available in two varieties: one for testing TTL logic circuits, and another for testing CMOS logic gates. More sophistocated (and expensive) probes incorporate switching to take account of the circuit differences.

Table 15.1

LOW	PULSE	HIGH	
OFF	OFF	ON	The probe is connected to a point with a steady logic 1.
ON	OFF	OFF	The probe is connected to a point with a steady logic 0.
OFF	OFF	OFF	The probe is connected to a point which is open circuit or an unidentified level*.
OFF	BLINK	OFF	Low frequency (relative) 1 MHz pulse chain with a mark to space ratio of approximately 1:1.
ON	BLINK	ON	High frequency (>1 MHz) pulse chain with a mark to space ratio of approximately 1:1.
OFF	BLINK	ON	Pulse train with a high mark to space ratio.
ON	BLINK	OFF	Pulse train with a low mark to space ratio.

* Logic levels for TTL are defined from manufacturers' specifications as logic 1 between the range of 2 V and 5 V, and logic 0 between 0.8 V and 0 V. Any voltage in between is therefore undefined.

A typical logic probe uses three different coloured LEDs as a display. One lights when the probe is connected to a point where the logic level is high, another when the logic level is low, and the third is used to indicate if the point under test has a changing logic level, and is usually labelled PULSE. Table 15.1 shows the typical indications likely to be obtained.

Index

Printed in the United States
26622LVS00002B/27

9 780750 634762